New York City Fire Department

FORCIBLE ENTRY REFERENCE GUIDE

TECHNIQUES AND PROCEDURES

INTRODUCTION

The objective of this manual is to provide the reader a comprehensive study of forcible entry. Although it cannot cover every aspect or technique of this demanding skill, it does cover those techniques that have proven to be successful for members of the FDNY.

The skill of forcible entry has been part of the fire service since its inception. The ingenuity and foresight of many talented people developed these techniques, which were then handed down through the generations of firefighters by "on-the-job training." It is our privilege to honor these people for providing the motivation and drive to put this material together. The goal of this book is not to take credit for these techniques, but to bring them all together for the benefit of the current and future members of the FDNY.

A program of training can be developed from using this manual, the forcible entry lock-board and the forcible entry training DVD that has been provided to the field.

Tools ....

- A Conventional Tools
- B Thru-the-Lock Tools
- C Hydraulic Tools
- D External Lock Tools
- E Power Tools
- F Specialty Tools
- G Modified Tools

Types of Locks ....

- Key-in-the-Knob Lock
- Tubular Lock
- Rim Lock
- Mortise Lock
- Magnetic Lock

Types of Doors ....

- Wood and Glass Panel Door
- Wood Door
- Metal Door
- Multi-Lock Door
- Tempered Glass Door

Types of Doors ....
- Aluminum Frame Glass Door
- Replacement Door
- Sliding Doors
- Pocket Doors

Additional Security Devices ....
- Sliding Bolt
- Static Bar
- Angle Iron
- Cylinder Guards
- Home-Made Locking Devices
- Lock Box

Conventional Forcible Entry ....
- Definition
- Entry Size-Up
- Steps for Forcing a Door
- Striking the Halligan Tool
- Alternate Methods to "GAP" a Door
- Halligan Tool Gets Stuck In a Door
- Drive The Lock Off the Door
- Angle Iron
- Narrow Hallway / Recessed Door / Tig
- Outward Opening Door
- Difficulty Gaining a Purchase
- Metal Strip On the Edge of Door

Hydraulic Forcible Entry Tools ....
- Steps For Forcing a Door
- Alternative Methods of Forcing a Door
- Angle Iron On Door
- Magnetic Lock
- Multi-Lock Door

Hinges ....
- Types of Hinges
- Forcing Hinges
- Batter the Door
- Standard Hinge
- Self-Closing Hinge
- Pin Hinge

Chocking the Door ....

Thru-the-Lock Entry ....
- Introduction
- Size-Up
- Key-in-the-Knob Lock
- Outward Swinging Door
- Tubular Lock
- Rim Locks
- Forcing a Rim Lock
- Special Type Rim Locks
- Forcing Special Rim Locks
- Mortise Locks
- Forcing a Mortise Lock

9

Thru-the-Lock Entry……………………………………………………………………

- Pivoting Dead Bolt
- Forcing the Pivoting Dead Bolt

Padlocks ……………………………………………………………………………

- Introduction
- Categories of Padlocks
- Padlock Size-Up
- Light Duty Padlocks
- Heavy Duty Padlocks
- Special Padlocks
- Gate Locks
- Associated Hardware
- Power Tool Procedures for Forcing Padlocks
- Other Tools for Forcing Padlocks

Roll-Down Security Gates ……………………………………………………

- Introduction
- Fire Ground Problems
- Types of Gates
- Sliding Scissor Gate
- Manual Roll-Down Gate
- Mechanical Roll-Down Gate
- Electric Roll-Down Gate
- Open Grill/Designer Gate
- Locking Devices
- Cutting the Roll-Down
- Additional Locks/Shields

Miscellaneous Security Problems ....

- Window Bars
- Window Gates
- Iron Gates
- Child Guard Gates
- Window/Door Barriers (HUD Windows/Doors)
- Plywood Covering Window/Door
- Warehousing
- Sidewalk Cellar Doors
- Bulkhead Doors

Tips and Techniques ....

- The Halligan Tool
- The Axe
- Modifying the Halligan Hook

Glossary of Terms

THE BEGINNING

THE BEGINNING

In the fire service, the term **"forcible entry"** is defined as the act of gaining entry into a building or occupancy via a door, window or even through a wall**, by the use of force.** Back through the years, the fire service has been charged with this responsibility of gaining entry into secured buildings and occupancies.

Forcible entry has always been a primary goal of the fire service. Over the years, the types of tools used for this purpose have evolved quite a bit. How many people out there can recall a "Callahan" door opener, the Buster Bar, Hale, Detroit or Pirsch door openers; the past generations of the current Rabbit Tool, or Hydra-Ram?

All of these tools had their place in the fire service. Technology and the imagination of skilled people designed lighter and more versatile tools. But the heart and soul of forcible entry usually comes down to two firefighters gaining entry through a door with a **"Set of Irons."**

The Claw Tool
Where did this term "irons" originate? According to Hugh Halligan, the man who invented the Halligan Tool, many years ago firefighters responded to a fire in a bank. The fire was started to cover a burglary. In their haste to leave with the money, the thieves left behind a tool used to gain entry into the bank. This tool was a heavy length of steel with a fork on one end and a claw on the other end. The firefighters who extinguished the fire reasoned that any device efficient to break into a bank would be ideal for fire fighting. The firefighters adopted it as their own forcible entry tool. Many believe this was the first tool specifically designed for forcible entry.

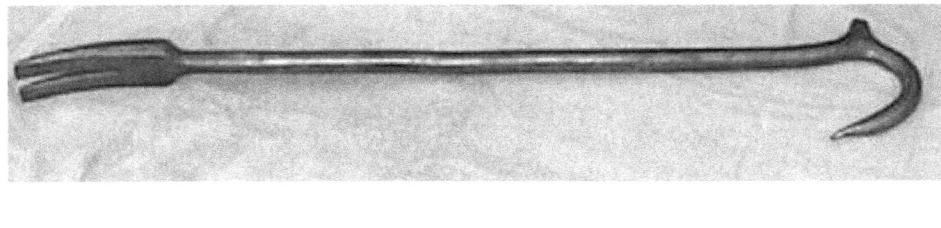

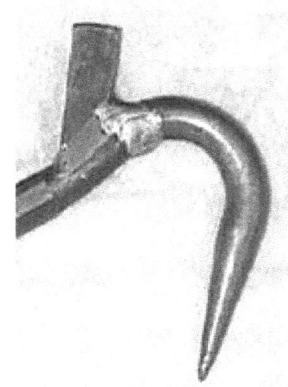
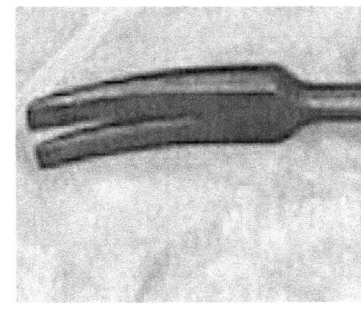

The Kelly Tool

Whether or not the story is true, the Claw Tool was used by the fire service for many years. Over the years, other tools were introduced to the fire service. Many were excellent, but were limited in their application. Then along came the Kelly Tool which received its name from the inventor, Captain John F. Kelly of H&L 163 (FDNY). His tool had a chisel at one end and an adz at the opposite end. The advantage of this tool over the Claw Tool was the striking area, which was in direct line with the bar. This tool was also known as the "Lock Breaker." It was designed as an alternate forcible entry tool to the Claw Tool.

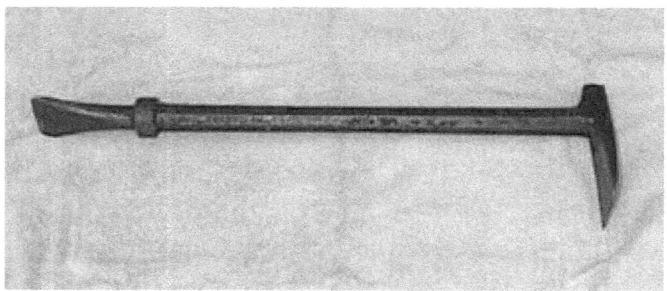

The "Irons"

The Claw Tool was still very popular with firefighters, especially its hook feature which gave quite a bit of leverage for forcing padlocks and scuttles. The Kelly Tool found its place by offering the straight drive of the adz and chisel. Together these tools could force just about any door or locking device. As the years went by, these tools became known as the "Irons" and were carried by the firefighter charged with the responsibility of forcible entry. Since they were usually carried connected by a short length of rope (hose strap) and hung over the shoulder of the member carrying them, he became known as the "Irons Man."

The Halligan Tool

Since these tools were quite heavy and unwieldy, the tools often "mastered the man." A lighter but equally efficient tool was needed. Along came Chief Hugh Halligan, FDNY, who took the design features of both tools and incorporated them into one hand tool. This tool had three driving heads. It was light (8 ¼ pounds) and incorporated the fork at one end and the adz and a slightly curved pike (instead of the claw) at the other end.

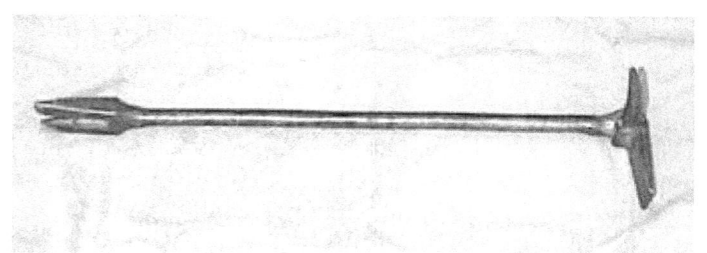

Chief Hugh Halligan With, The "HALLIGAN TOOL"

The Ziamatic Tool
In the early sixties, fire duty began to increase in New York City. At that time, the only tools the FDNY was issuing to its units were the Claw and Kelly Tools. (Folklore has it that Chief Halligan would not sell his tool to New York City.) Today we have many variations to the Halligan Tool. Some are even better than the original.

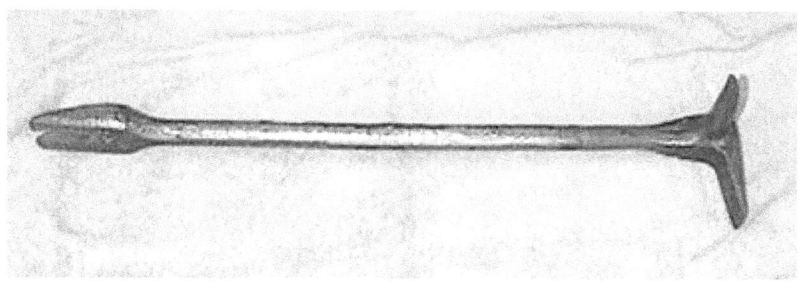

Some manufacturers took this good tool and made it better, others just copied the original design. One such company, the **Ziamatic Tool Company** began reproducing a similar tool. This was one of the many variations to the original Halligan Tool. It was quickly purchased by the New York City Fire Department to augment their limited supply of forcible entry tools.

The Pro-Bar
This tool has quickly become the FDNY's primary forcible entry tool. Many young firefighters consider this "the Halligan Tool," but it is just one of many copies of the original design.

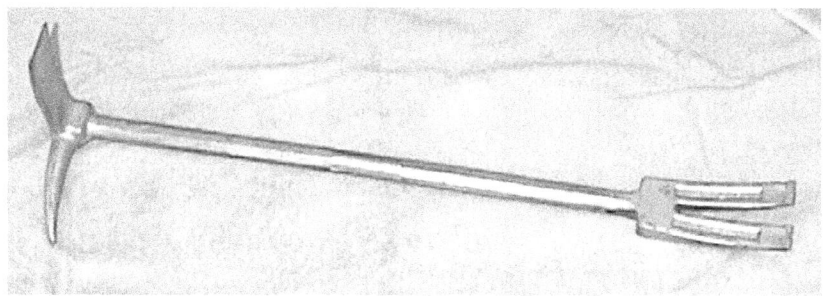

This particular tool was also the brainchild of former New York City firefighters. They took the original tool and combined the better features of the Claw and Kelly into a better designed tool; hence the "Pro-Bar."

Comparing the "Original" with the "Pro-Bar"
The original Halligan Tool was unique to forcible entry. Combining many tools into one compact, hand tool took a keen mind. Chief Halligan did indeed make a revolutionary tool. However, there were some shortcomings with the original design.

The blunt fork and short narrow adz may have been effective in the early years, but due to new security technology, the original tool became inefficient. A simple modification to the original design proved to be quite effective. To this day, the modifications produced have proven to be most effective for a hand tool.

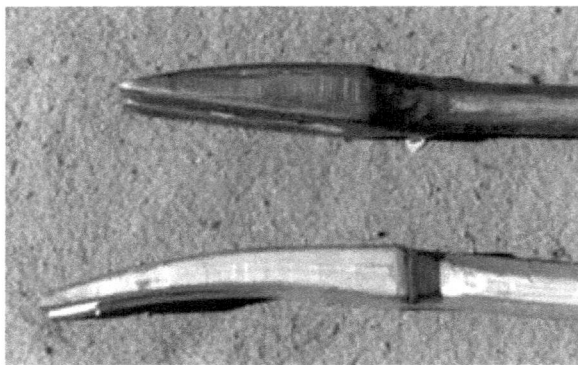

The fire service had been challenged to find other methods of gaining entry. At times it may require a different technique, more skill and stronger tools to accomplish this. This manual will attempt to:

- Outline principles, methods and techniques that will insure the effective use of forcible entry in training and in fire operations.
- Promote uniformity in training.
- Provide a handbook for the teaching and learning of forcible entry.

RESPONSIBILITY

RESPONSIBILITY

Again it is important to understand that the fire cannot be extinguished, searches cannot be made, and extension of fire cannot be checked until entry is made. The fire fighter assigned the job of gaining entry is given that responsibility. To accomplish this task, there are an assortment of tools and techniques, which this text will introduce to you. Some techniques are basic, others are more difficult, but all are achievable.

Proficiency:
Why all firefighters should be proficient in the basic forcible entry skills.
- **The need for speed in gaining entry.** It is important to realize that most fire and emergency operations start at the front door or main entrance. Before any tactical moves can be made, e.g. search, rescue or the stretching of a hand line to the seat of the fire, the entry door has to be opened.
- **Reduce damage resulting in improper techniques.** Most people given tools can gain entry. A door can be "battered" down with an axe (the movie version). However, until we take into account what is behind that door, we want to ensure the door's integrity. Why destroy a perfectly good door for a non-fire emergency? With the proper training, most firefighters will be able to open a door with minimal damage.
- **Professionalism.** This is the benchmark of a good firefighter. The firefighter represents the department and ultimately the city or hamlet. Pride in our work will reflect pride in the department. By reducing the damage to a minimum we ensure the safety of the people we serve. Remember that when we leave the fire scene, the doors we destroy leave the occupants vulnerable to further loss from vandalism. The people we are sworn to serve rely on our good judgement.

Jimmying A Door:
The old technique of **"jimmying a door"** (the spreading of the door away from the jamb without damaging the lock) can seldom be accomplished today. This is due to stronger doors, more formidable locks and multiple locks on a single door.

The primary motivation should be professionalism. As a firefighter, you have an obligation to get the job done **safely, efficiently and with the least amount of damage.** At times, brute force must be combined with skill, technique and knowledge. You control that action.

For situations such as: water leaks, steam leaks, lock-ins, etc, consider the least damaging means of gaining entry. In some instances, you may be able to enter through a window or by using a "Thru-the-Lock method of entry. Always **use common sense** when forcing your way into any premises; you never know what is behind that door or window.

You must also consider what will happen once your job is done. Who will provide security for the occupancy after you leave?
In order to become proficient in the skill of forcible entry, you should have a mixture of:

Hands on training- this is the primary way to sharpen your skills.

Experience- by going to fires and emergencies and actually "forcing the door."

Knowledge- may be gained by experience, reading, observing, attending training seminars and also by exchanging information and ideas with other firefighters.

Finally, using some **common sense** and trusting your instincts; they are usually correct.

"Why Are You There?"

What are the reasons for entry? Is it a **Tactical Response?** That is, for a fire and/or life-threatening emergency, or is it a **Routine Response** for a non-life-threatening emergency? In either situation, control, speed and effectiveness of access to the area of operations will justify the amount of damage done by the firefighter. Remember, the goal is to: **save life, extinguish fire and control all hazards.**

Size-Up:

This is the ongoing evaluation of the problems confronted within a fire situation.

As you get off the apparatus, you should be asking the following questions:
- Where is the fire?
- How many floors?
- What type of occupancy?
- What type of building?

Size-up starts with the receipt of an alarm and continues until the fire is under control. This process may be carried out many times and by many different individuals during a fire or an emergency.

In conducting a size-up we should consider the following:

- **Occupancy:** Knowing you are responding to a residential or commercial occupancy will help determine the type of doors and locks you may encounter. This will help determine what specialized tools may be required.

- **Door:** Knowledge of the type of door and its components may guide you as to proper tool placement and method of entry. This would include:
    1. **Direction of door opening:** most **residential doors** open into the occupancy. They are considered **inward opening** (away from you). Whereas in **commercial occupancies,** the door opens out of the occupancy. They are considered **outward opening** (toward you).
    2. **Door Frame:** A structural case or boarder into which a door is hung. Also referred to as a **Door Buck**, **Door Jamb** or simply, the "**Frame**." They can be made of metal or wood.
    3. **Hinges:** There are many types of hinges used today. The types we discuss here will be known as (a) standard, (b) self-closing, and (c) pin type.
    4. **Replacement Door:** A new pre-hung door and jamb installed into an **existing** doorframe.

- **Locks:** To determine the degree of difficulty in forcible entry you should have a working knowledge of the **various types of locks** as well as a basic understanding of how they operate and how they are installed. One should also take notice of how many locks are present and where they are located on the door.

- And finally, you should always **TRY THE DOOR KNOB** - "is the door open?"

TOOLS

TOOLS

The success of any job resides in the knowledge of the tools and their correct application. Listed here, within categories, are many of the tools used in forcible entry:

Conventional Tools

- Axe (6 and 8 pound)
- Halligan Tool
- Maul (10 pound)
- Halligan Hook (steel shaft)

Thru-the-Lock Tools

- K-Tool and Key Tools
- Lock Puller (Officer's Tool)
- Shove Tool
- Vice Grips (may be used for Padlocks, Thru-the-Lock)

Hydraulic Tools

- Hydra-Ram
- Rabbit Tool

External Lock Tools

- Bam-Bam Tool
- Duckbill Lock Breaker
- Bolt Cutter
- Pipe Wrench with Cheater Bar

Power Tools

- Power Saw
- Cordless Drill/Cordless Sawzall

Specialty Tools (Limited use)

- Torch
- Battering Ram
- Vice Grips (may be used for Padlocks, Thru-the-Lock)

Modified Tools – Standard tools/devices that have been modified for use in the fire service.

- Channel Lock Pliers
- Key Tools
- Padlock Tool
- 8-Pound Axe

The following are brief descriptions and reasons we chose the above tools for **Forcible Entry**. There may be firefighters that have a different approach or use different tools to accomplish the same end, but these are the tools we have used and are most familiar with.

CONVENTIONAL TOOLS

Axe (6 and 8 pound): This should be a **FLAT HEAD** type axe and not a pike head axe. The purpose of this axe is to drive (SET) the Halligan Tool. There are two sizes available and choice is up to the unit. The 6-pound axe can easily be "**married**" to the Halligan Tool for carrying.

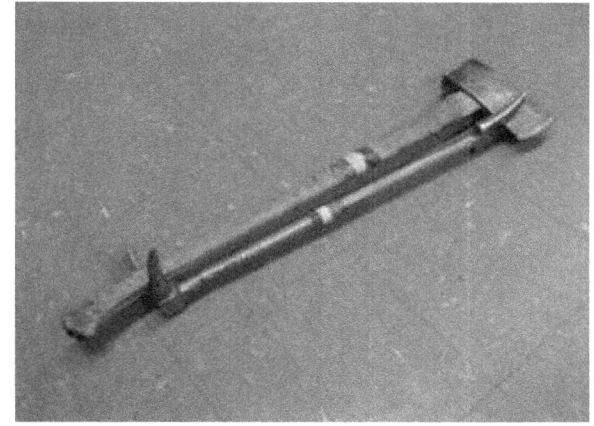

The 8-pound axe may not "marry" up due to its blade size. However, notching the blade can modify this. (See "Tips and Techniques" Chapter 16.) The 8-pound axe will deliver more power to the Halligan Tool. Either axe should be "dressed," e.g. the striking part of the axe should be filed and kept square. Avoid having the crown of the axe from "mushrooming" over.

The axe with the Halligan Tool form the "Irons" which are the basic forcible entry tools. The axe can also be used to:

- "Chock open" the door.

- Be a backstop for the Halligan or hydraulic tool (Hydra-Ram).

- Hold the purchase when repositioning the Halligan Tool.

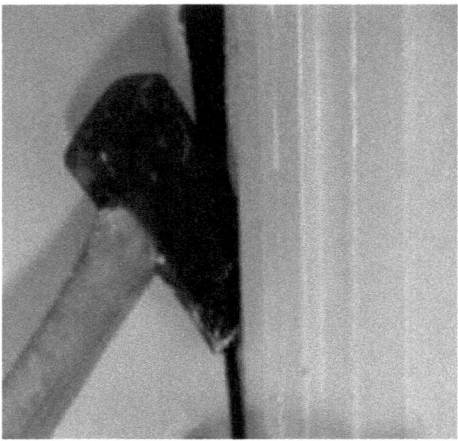

Halligan Tool: There are many models of this popular tool. The one illustrated here is approximately thirty inches long with a beveled fork, a tapered adz and pike. For more details refer to "Conventional Forcible Entry," Chapter 8.

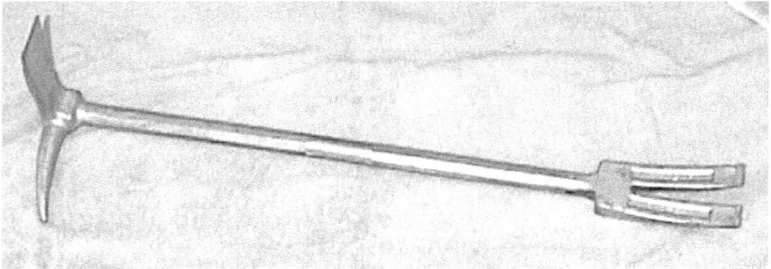

Pro-Bar Halligan Tool

Notch in Axe Blade
By filing a notch into an 8-pound axe, a Halligan Tool may be "married" to it allowing the member to carry both tools in one hand.

There are straps that are sold commercially to join the two tools, but it just adds to more equipment to carry and be responsible for.

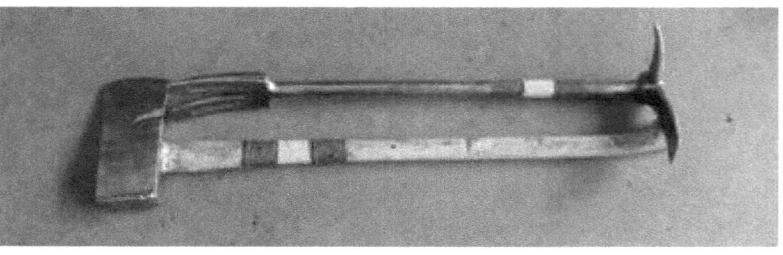

Maintenance of the Irons

Proper maintenance of tools and equipment is the first step in tool safety. Tools should be inspected and cleaned on a regular basis. Always check for wear and damage. If equipment is found damaged it should be removed from service until repaired or replaced. Proper care of forcible entry tools will increase their serviceability.

Metal parts

- Remove any dirt or rust with steel wool or emery cloth.
- Use a metal file to maintain the proper profile and cutting edge.
- Sharpen edges and remove any burrs with a file.
- Do not keep the blade edge too sharp as this may cause it to chip when in use.
- Do not grind the blade as this may overheat the metal and cause it to lose the temper.
- Do not paint the metal parts, but keep them lightly oiled if desired.
- Never apply oil to the striking surface of a striking tool (axe or Halligan).
- "Dress" the edges to keep square and free of burrs which may splinter off when striking tool.

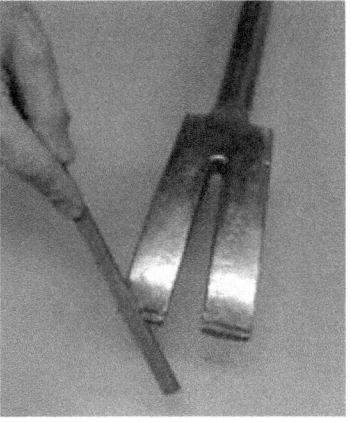 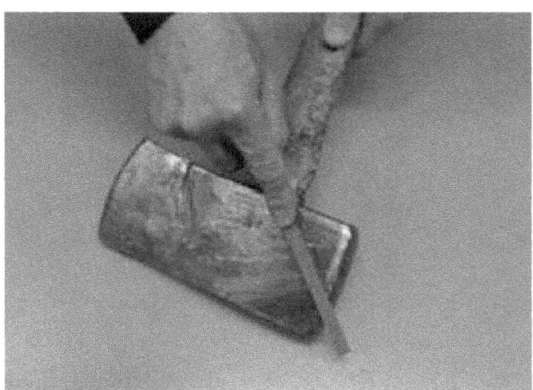

Wood and Fiberglass Handles

- Clean with soap and water; rinse and dry completely.
- Check for damage and sand off any splinters.
- Do not paint or varnish the handle. A small band of paint or brand may be used to identify the tool.
- Ensure the head of the tool is securely fastened.
- Use tape to mark off a narrow stripe on handle to identify unit.

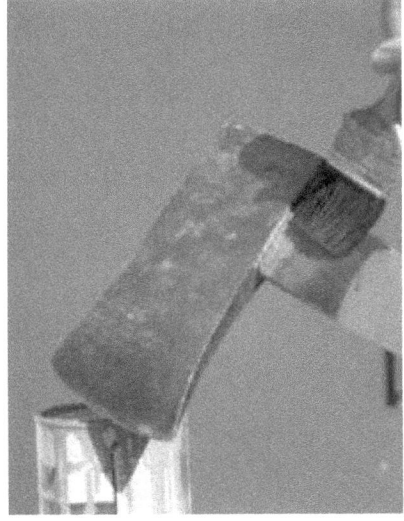

Maul (10 pound): This tool comes in a variety of sizes, but the most common and versatile is the 10 pound model. This tool may be used in place of the axe to form the "Irons." Other uses would be to "batter" a door or to remove cinder block from a window or door of a vacant and sealed occupancy.

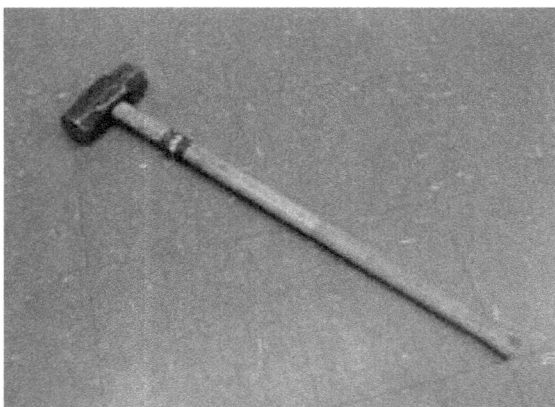

Halligan Hook (steel shaft): This tool is a six foot, steel shaft hook, with a distinct shaped head and is commonly referred to as a "Halligan Hook."

These are primarily "pulling tools," e.g. for pulling ceilings. For entry, the steel shaft can be used to set the Halligan Tool into a tight doorframe (such as a bulkhead type door) by "toeing" on the end of the shaft and driving the Halligan Tool with the shaft.

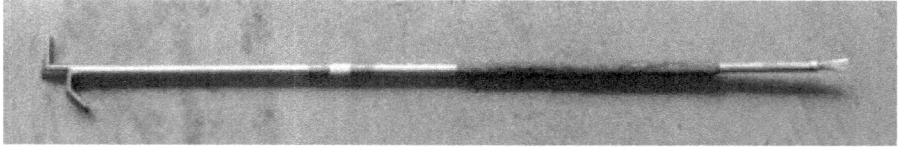

Metal Halligan Hook

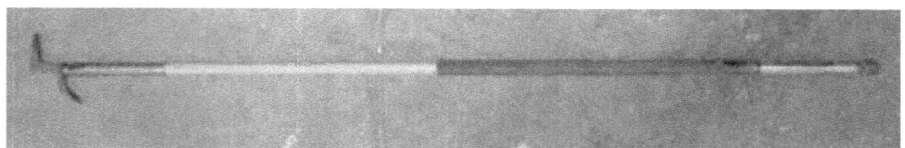

Fiberglass Halligan Hook

THRU-THE-LOCK TOOLS

K-Tool: This tool was developed for pulling a lock cylinder (Thru-the-Lock entry) on a door. It is used with an axe and Halligan Tool.

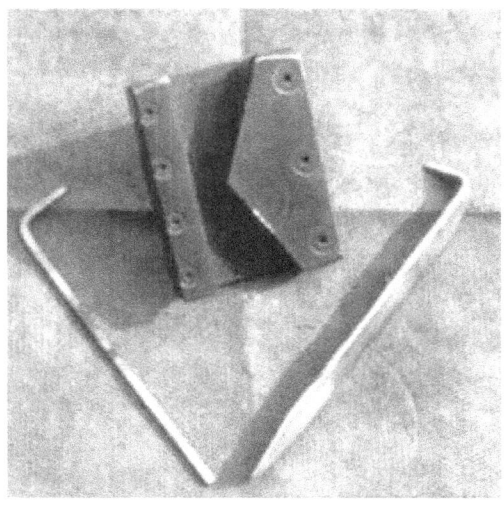 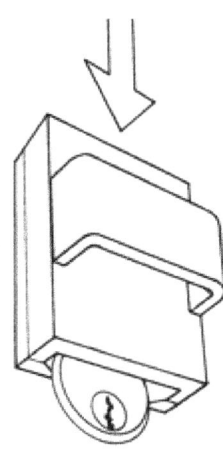

The K-Tool is forced behind the ring and face of the cylinder until the wedging blades take a bite into the cylinder body. Light blows with the axe set the K-Tool.

The Halligan Tool's adz is placed into the slot on the face of the K-Tool and pried upwards, pulling the cylinder from the door.

Lock Puller: It is a device developed from a modified nail puller called the "Sunilla Tool," named after its inventor, Captain Sunilla (FDNY). This is one of the first tools designed to pull cylinders out of locks. It is also useful for opening automobile trunks.

There are various designs and shapes being sold throughout the country. They have a wide variety of names and uses. In certain parts of the country, this tool may be carried by the officer (hence the Officer's Tool).

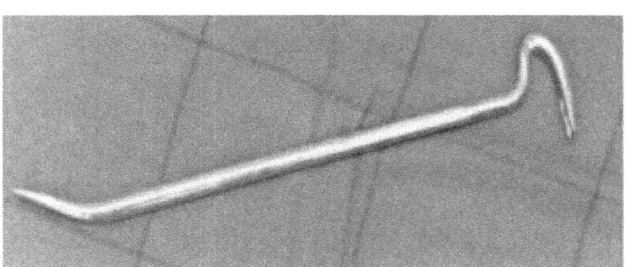

Sunilla Tool

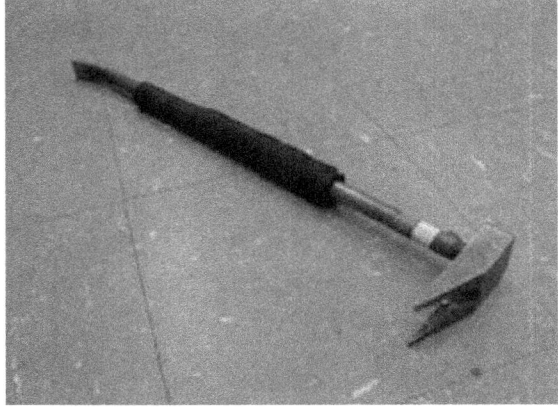

Officer's Tool FDNY

Shove Tool: It started out as a device to slip the latch on a door. Many were first produced by enterprising firefighters from old hand saw blades or similar materials. Today, tool manufacturers are producing them. It is flexible, 10 gauge sheet steel, approximately eight inches long by one and half inches wide. The device is slid between the door and the doorframe above the spring latch. Once the "hook" end catches the latch, the tool is pulled toward the operator which depresses the spring latch opening the door. **It only works on outward swinging doors.**

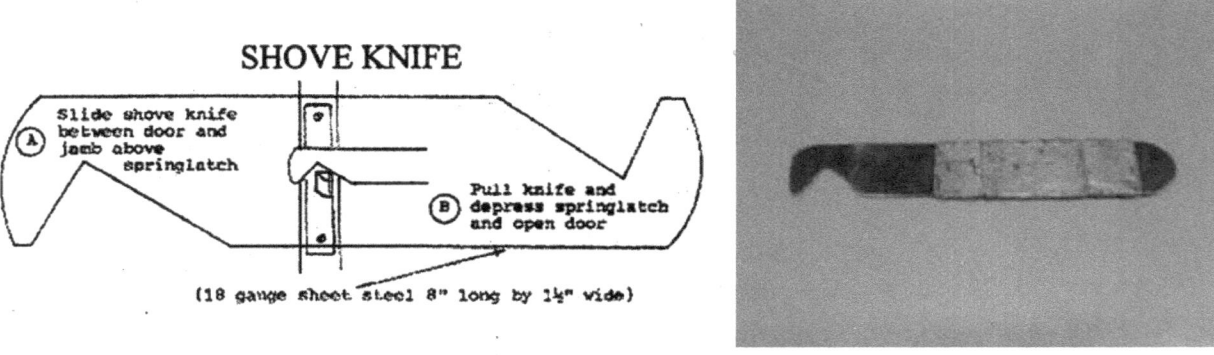

Vice Grips: A very useful tool for any firefighter's tool box. This locking pliers can be used to "unscrew" a mortise lock cylinder from the lock housing or to simply hold a padlock while it is being cut with a power saw.

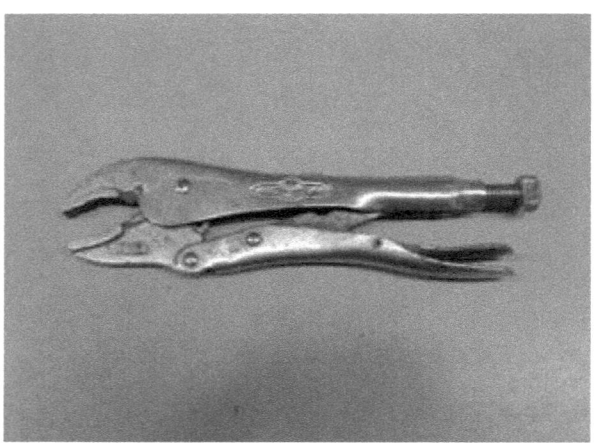
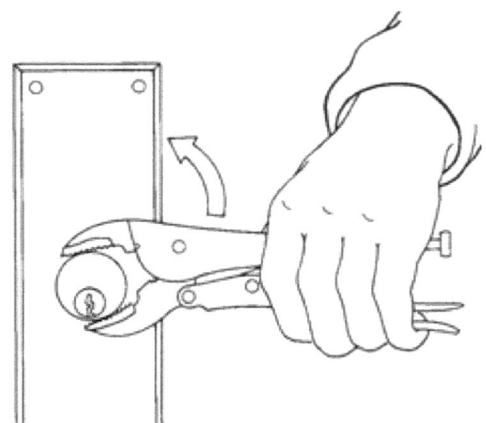

HYDRAULIC TOOLS

These tools are used for forcing inward swinging doors. They work best on doors mounted in metal frames. They have also been used to force sliding doors found on passenger elevators. More information will be found in Chapter 9, Hydraulic Forcible Entry Tools.

Rabbit Tool: One of the first hydraulic forcible entry tools to be introduced in the FDNY. It is a two-piece unit connected by a high-pressure hose. The large jaw will spread force over a greater area. It exerts over four tons of force with a jaw spread of approximately six inches. The weight of the tool is 25 lbs. The pump is designed to be operated in the horizontal position, but may be used vertically if the hose is facing down.

Hydra-Ram: The second generation hydraulic forcible entry tool to be introduced to FDNY. This is a one-piece unit weighting 12 lbs. The maximum force the tool will exert is five tons with a jaw spread of approximately four inches.

EXTERNAL LOCK TOOLS

Bam-Bam Tool: Also known as a "Slap Hammer." This tool was primarily used in body shops to pull dents out of automobiles. It has proven quite successful in pulling lock cylinders from many padlocks. It requires a good quality self-tapping screw. More on this in "Padlocks," Chapter 13.

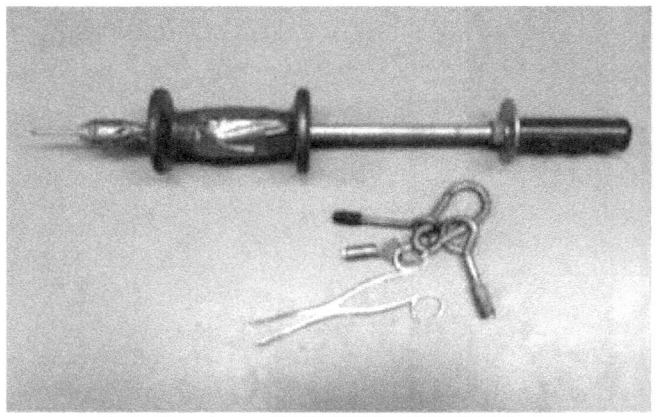

Duckbill Lock Breaker: Another tool that was modified from a laborer tool, the "Pick-Axe." It is used to drive the body of the padlock off the shackle. The long tapered head is placed into the shackle of the padlock and driven down with a flat head axe, maul or even the Halligan Tool.

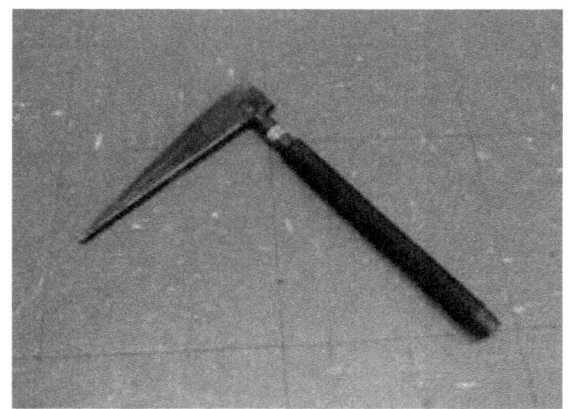

Bolt Cutter: Another tool used for cutting hasps, light-duty padlocks and chains. It is limited by the opening spread of the blades. It is not recommended for cutting case-hardened shackles since that may damage the cutting blades. If possible when cutting, try to cut the staple holding the padlock. If you have to cut the padlock, cut both sides of the shackle.

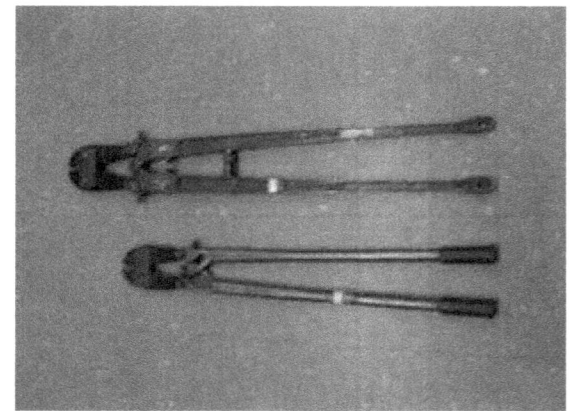

EXTERNAL LOCK TOOLS

Pipe Wrench With a Cheater Bar: This is a large pipe wrench with a piece of pipe over the handle to give the operator more leverage. With a little initiative from the user this tool can be modified to gain additional leverage.

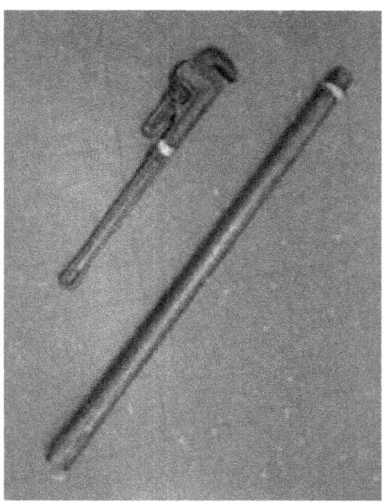
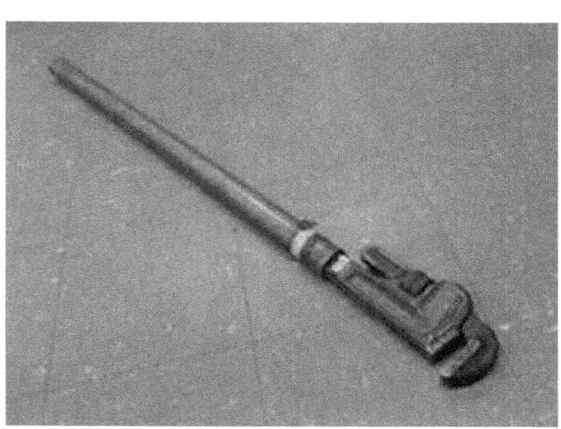

POWER TOOLS

Saw: The Power Saw improves forcible entry efficiency by facilitating cutting operations at fires, especially where roll-down security gates are present. These saws come in a variety of models. They require a metal cutting blade when cutting padlocks and/or roll-down security gates. The saw is usually run at low Rpm's until a groove is made in the metal, the power is then increased to maximum speed to complete the cut. More in Chapter 14, Roll-Down Security Gates.

POWER TOOLS

Cordless Drill: Relatively new to the fire service, it operates off of a battery. A method of Thru-the-Lock entry which causes minimal damage to the door. It is a convenient tool for gaining entry into high-rise office buildings.

Cordless Sawzall: Relatively new to the fire service, it operates off of a battery. This tool is quickly becoming multi-versatile. Not only is it good for removing gates and bars, but it is also used in vehicle extrication.

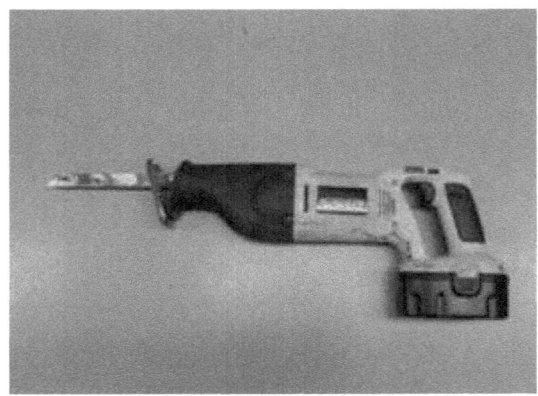

SPECIALTY TOOLS (Limited Use)

Cutting Torch: Many torches used today utilize Mapp Gas and Oxygen for cutting steel and iron for the purpose of entry or rescue. This is a safer alternative to Oxy-Acetylene for cutting gates and locks.

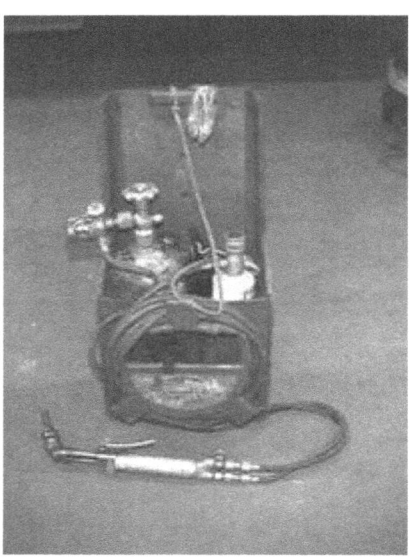

SPECIALTY TOOLS (Limited Use)

Battering Ram: There are quite a few models of this device used for breaching walls and forcing doors. It usually has handles on both sides and may be used by one or two firefighters. At one time this was used for forcible entry, today it has limited use in breaching walls.

MODIFIED TOOLS

Standard tools and/or devices that have been modified for use in the fire service. Some of the many types out there are shown below:

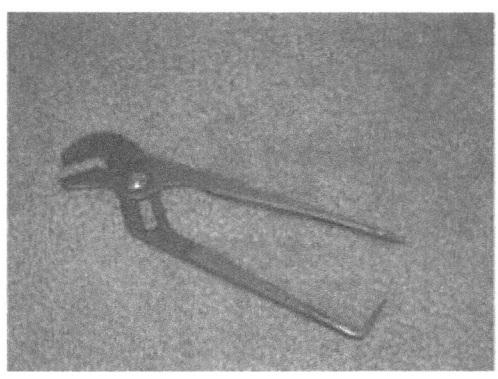

Channel Lock Pliers: Modified commercial Channel Locks into two Key Tools for the Rim and Mortise type locks.

Key Tools: Eyebolts and standard 10-Penney nails modified as Key Tools.

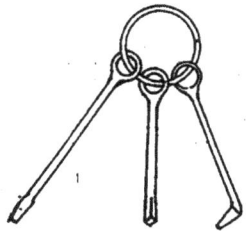

Eyebolts

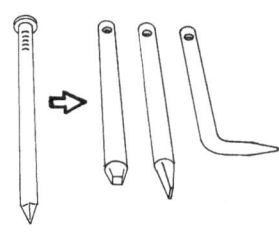

10-Penny Nails

MODIFIED TOOLS

Padlock Key Tool: Field modified device using a threaded "eye" bolt welded to the "pin" from a previous pulled cylinder. Works mostly with the "American Series 2000" padlock.

 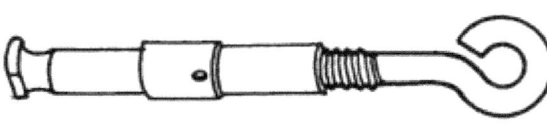

NOTE: This is not a complete list, as new tools and equipment are constantly being introduced.

TYPES OF LOCKS

TYPES OF LOCKS

KEY-IN-THE-KNOB LOCK - As the name implies, the locking mechanism is part of the knob. These locks are found on both residential and commercial doors.

 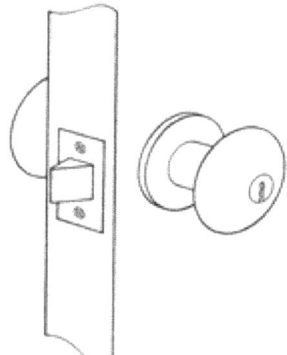 

TUBULAR DEAD BOLT - This is a very popular locking device. It may be single or double key activated. It is a cross between a mortise lock, rim lock and a key-in-the-knob lock.

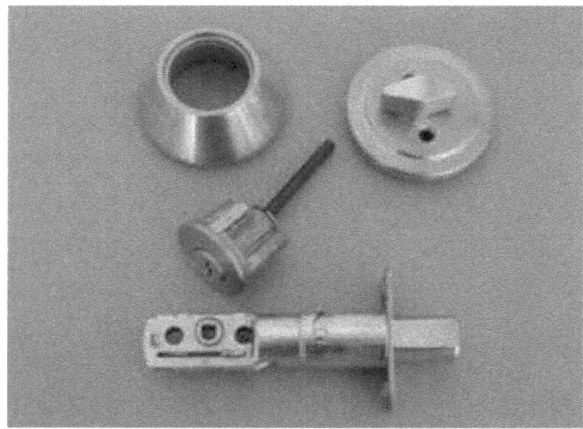

RIM LOCKS - These locks are usually installed as an **add-on lock.** They are installed on the **inside surface of the door** (with the cylinder extended through the door). Only the cylinder is visible from the outside of the door.

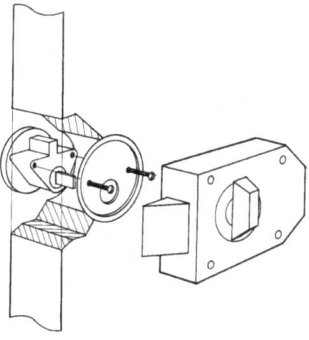
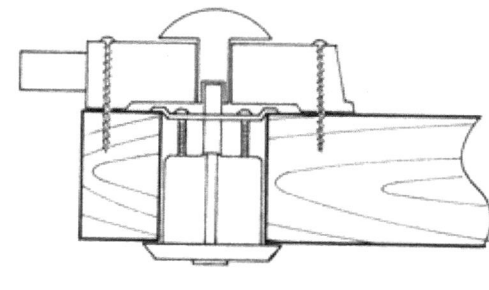

Deadbolt - Unlike a spring latch, this device must be manually thrown to engage the bolt into the keeper. With the bolt extended, this lock cannot be engaged by slamming the door.

Night Latch - The latch is beveled to allow the door to be slammed shut. Some of these spring latches have an inside button to prevent the latch from returning within the lock, e.g. sliding open.

Vertical Dead Bolt (Segal Lock) - This rim lock has a bolt which drops down and through the keeper. This device must also be manually engaged. It is a "jimmy" proof lock.

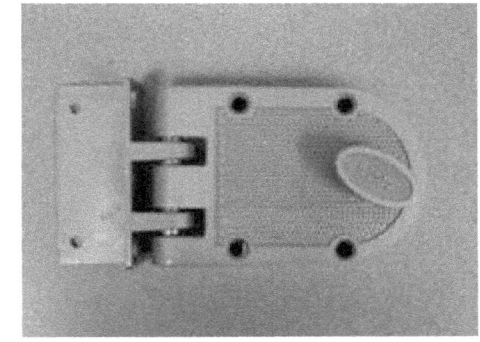

MORTISE LOCKS - Are designed and manufactured to fit into a cavity in the edge of either a metal or solid wood door. They have a solid, threaded key cylinder, which is secured in place by set-screws. The two most common types are; Mortise/Latch Key and Mortise/Door Knob (see below).

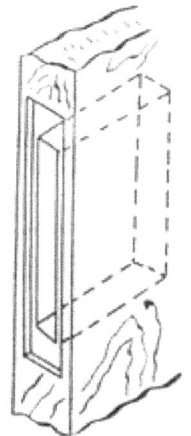 

DEAD BOLT AND LATCH - One of the most popular locks in use today. It contains both a latch and a bolt in a single unit. It is distinguishable by the proximity of the lock cylinder and a door knob or latchkey. Below are examples of this type of lock.

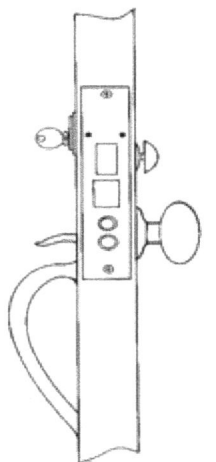  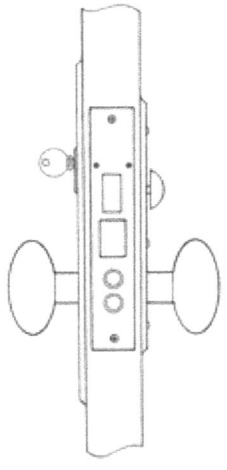

Mortise / Latch Key   **Deadbolt And Latch**   **Mortise / Door Knob**

MAGNETIC LOCK – A relatively new locking device that has been incorporated into occupancies for added security.

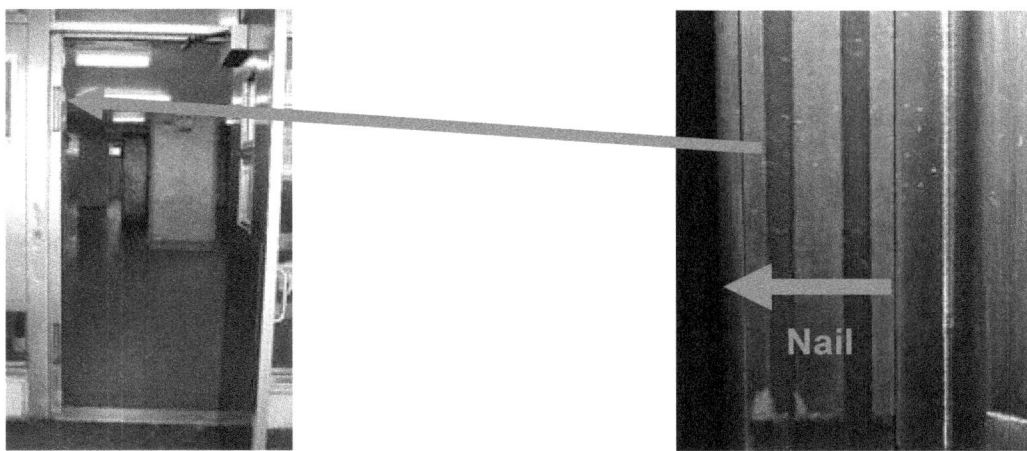

Note: Placing a common 8-10 penny nail over the magnet will prevent the door from re-locking.

TYPES OF DOORS

TYPES OF DOORS

WOOD AND GLASS PANEL DOOR - This was a very popular door in older buildings. It provided light to the public hall in multiple dwellings. The original plain glass panels were changed to wire glass. Some wood and glass doors may contain plate glass. Today these are found in Brownstones and some older "Mom and Pop" stores.

Note: Plate glass may be quite dangerous. When broken, it may fall in large sharp pieces. These pieces have significant weight and force to cause serious cuts or stabbing and dismembering injuries.

WOOD DOOR - There are two types of wood doors; Hollow Core and Solid Core.

Hollow Core: Made up of an assembly of wood strips formed into a grid. These strips are glued together within the frame forming a stiff and strong core. Over this framework and grid are layers of plywood veneer paneling.

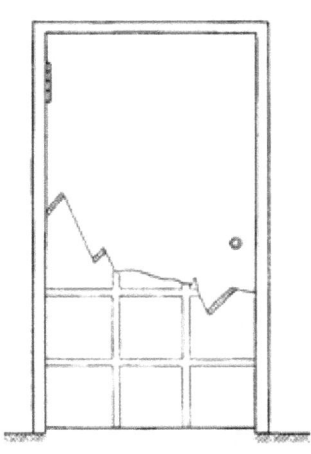 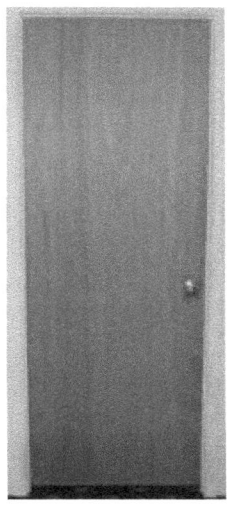 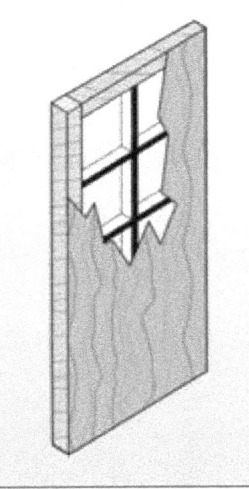

WOOD DOOR

Solid Core: The entire core of the door is constructed of solid material such as tongue and groove boards that are glued within the frame. Other solid core doors may be filled with a compressed material that is fire retarded. In either case, the door is sided with a plywood veneer covering.

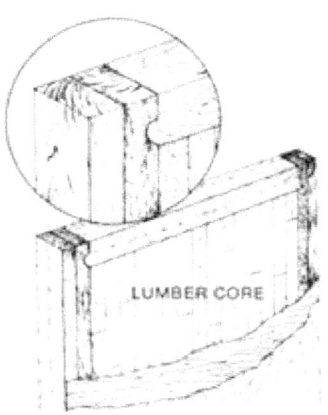
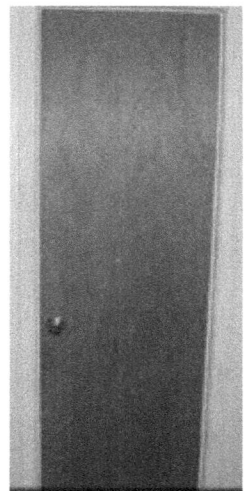
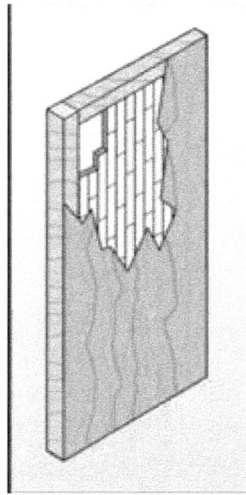

KALAMEINE DOOR

The main problem with a wood door, especially in multiple dwellings, was the "burn-through" time. To overcome this problem and to increase the burn-through time, these doors were covered with metal. They were known as "**Kalameine Doors.**"

METAL DOOR (Project Doors)

Constructed of metal, these doors are usually set in hollow or filled metal doorframes. When set in a masonry wall, as well as a metal frame, they are quite formidable and will hold back considerable fire. Today a metal door is quite common even in private dwellings.

MULTI-LOCK DOOR

One of the most advanced locking systems available that utilizes a key and multiple bolts and keepers. Built **into the door**, are four rods, which extend out from a keyway toward all **four** edges of the door. The throw of each rod is approximately an inch into the frame. It is designed to prevent any rod from moving separately. Originally built as a deterrent against terrorism, it is used today in occupancies where security is very important.

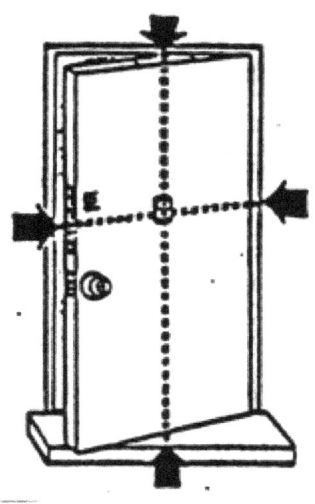
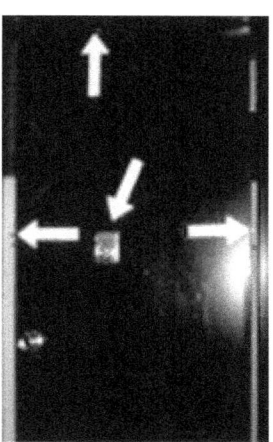

MULTI-LOCK (Add-On)

With the popularity of the multi-lock door came this less expensive version which is mounted on the inside surface of the door. Similar to a **Rim** lock, attached to the inside of the door are four bars, which extend out from a keyway toward all **four** edges of the door. The throw of each bar is approximately one inch into the frame or keeper. It is designed to prevent any rod from moving separately. When properly installed, it is equally as effective as the Multi-Lock door.

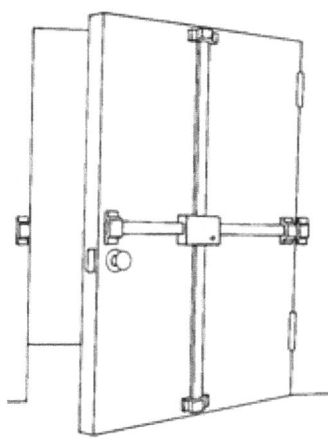

TEMPERED GLASS DOOR

Distinguishable by the lack of a full frame with little or no trim. The door handle is usually mounted through the glass. The lock may be installed in either the top or bottom stile **usually the bottom one.** Commonly known as a "Glass Door."

The breaking characteristics of Tempered Glass are quite different than ordinary Plate Glass. This is due to the heat treatment given to the glass during tempering. This results in high-tension stress in the center of the glass and high compression stress in the exterior surfaces. These tension and compression stresses balance each other. The heat treatment also increases the strength and flexibility as well as the resistance to shock, pressure and temperature increases.

Approximately four times stronger than plate glass, when broken, tempered glass disintegrates into relatively small pieces.

ALUMINUM FRAME GLASS DOOR

These are the most popular doors in commercial occupancies, especially the taxpayer type. It is not uncommon to have the plate glass replaced with tempered glass, lexon or plexi-glass in some areas.

REPLACEMENT DOOR

A relatively new situation where the existing doorframe is covered with a pre-hung steel door and a new metal pre-assembled doorframe is attached to it. This replacement assembly is screwed into the old frame. It may sound simple, but it is difficult to recognize and placement of the forcible entry tools must be between the **new doorframe** and the **door** and **not** between the **old frame** and the **replacement doorframe.** The door with the frame is laid in over the existing frame.

 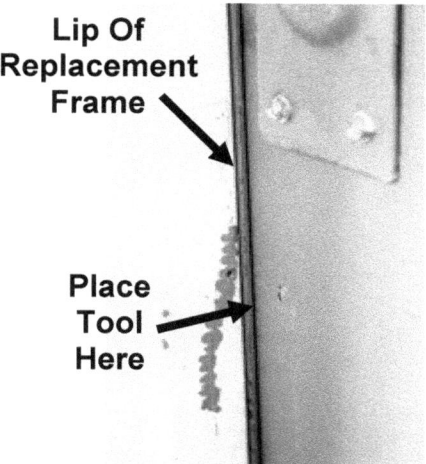 

Replacement Door   **Lip of Replacement Frame**   **Replacement Door Frame**

SLIDING DOORS

These doors may travel either to the right or left of their opening or in the same plane.
Sliding doors are usually supported upon a metal track and their side movement is made easier by small rollers or guide wheels.
A bar may be placed between the fixed frame and the door or in the track to prevent unlawful entry.

POCKET DOORS

These doors were quite popular many years ago and have found resurgence in today's construction.
They are interior sliding doors that slide into a partition or wall when pushed open and may be referred to as **"pocket doors."**
These doors may be forced similar to a swing door, except that they must be pried straight backward from the lock.
A major drawback is the voids they create in a fire situation.

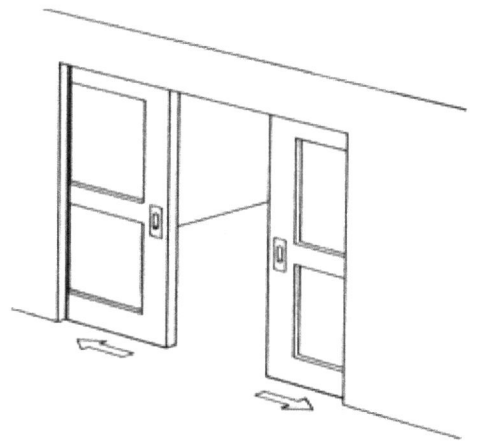

ADDITIONAL SECURITY DEVICES

ADDITIONAL SECURITY DEVICES

SLIDING BOLT

A device that travels in a track, which locks into a recessed hole or hardware. Padlocks may pass through rear of bolt and make the bolt secure. These slide bolts may be made of case-hardened steel. They are installed with screws or carriage bolts, which may be exposed or guarded.

STATIC BAR

A fastening device that can be mounted across the door at any point. Generally they are in pairs. The bars are held in place by brackets, which may be fastened to the doorframe.

Outside View

NOTE: With the Sliding Bolt and Static Bar in place, you know the occupants did not exit through that door. There is either another means of egress or the occupants are still inside. Static bars in place may not be visible from the outside.

ANGLE IRON

A device secured to the door and occasionally the doorframe. It can be found on both inward swinging doors (away from you) and outward swinging doors (toward you). It may be partial or run the full vertical length of door. It represents another form of security which may be added to an occupancy.

SHIELDED ANGLE IRON

A device that is mounted to both the door and the frame and inter-locks on itself. It may be partial or run the full vertical length of the door. It is two separate pieces mounted, one to each surface. By adding this inter-locking piece of angle iron additional security is added to the occupancy.

NOTE: The arrow points out a lock cylinder located NEAR THE BOTTOM of the door. This simple but ingenious set up prevents most "push-in" forcible entries.

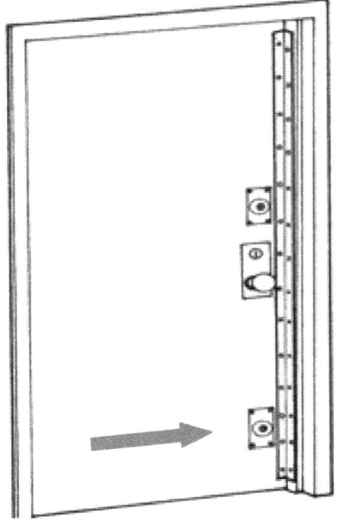

CYLINDER GUARDS

A raised rectangular metal plate over the lock cylinder that is held in place with four carriage bolts. These bolts may be exposed or hidden in the body of the guard.

Black

Heavy Duty

Brass

HOME MADE LOCKING DEVICE

A recent ingenious method of securing a door has started to appear in multiple dwellings. This is a home made modification of a "chain lock."

Here the occupant bolts a length of heavy chain to the inside of his door. (The Chain is similar to that which secures motorcycles.) Generally the carriage bolt and washer are secured approximately one foot or less from the edge of the door and about one-foot above the doorknob. The other piece of chain, similar in size and strength, is attached to the doorframe.

Joining the two pieces is a heavy-duty padlock.

What makes this device so ingenious is its simplicity and effectiveness. Since the carriage bolt may be overlooked, the forcible entry team will force the door, and then be confronted with a heavy-duty chain and lock which continues to secure the door.

Most people know a chain and lock can be quite formidable, especially if not under tension. Add to this the products of the fire venting out through the opening created by the initial forcible entry. Now the team must remove the chain and lock under much worse conditions.

Suggestions: In your size-up of the door, check for the presence of a bolt head in the door. If you suspect this is the chain lock, drive the bolt head through the door **BEFORE** forcing the door. This can be done with the pike of the Halligan Tool and sharp blows delivered with the axe or maul. **Size-up is very important**. If you miss the bolt head on the door, entry may be delayed.

HOME MADE LOCKING DEVICE

If the bolt is missed, and the door is forced open, lock the fork of the Halligan Tool around the chain at the frame side and try to pull it out of the frame. While doing this, maintain pressure on the door in the open position.

If fire emits from the open door, close the door until a charged line is in position, then continue as above.

This is not a simple operation. If the chain is bolted through the frame or secured with more than a single bolt, a forcible entry saw may have to be used. In this case, a delay will cause the fire to accelerate.

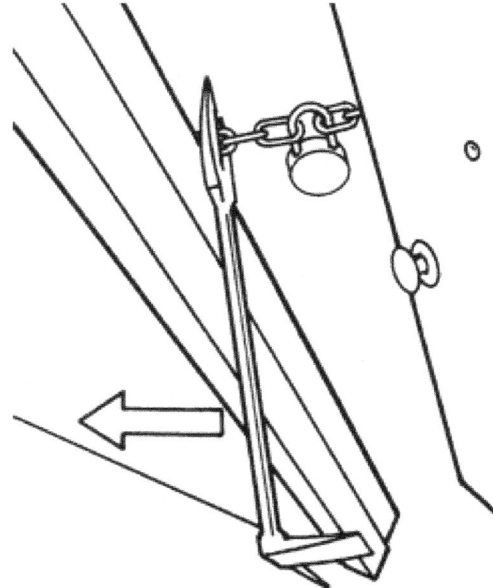

Lock Halligan Into Chain And Attempt To Pry Chain From Frame.

LOCK BOX

This method of security not only protects property, but limits damage to many locks and doors. It is a system of storing all necessary keys to the building or occupancy in a box that is mounted in a high visibility location, usually in the front of the building.

Only the local fire department carries the master key, which cannot be duplicated. This approach provides a high degree of security and eliminates the need to carry many individual keys. This has been around for many years.

Proper mounting is the responsibility of the property owner. The local fire department should indicate the desired location for mounting. They would inspect the completed installation and place the building keys inside, then lock the box with the department's master key.

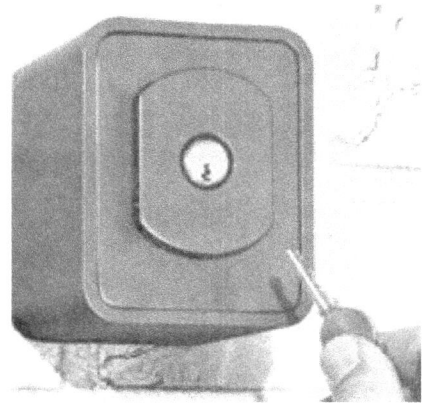

CONVENTIONAL FORCIBLE ENTRY

CONVENTIONAL FORCIBLE ENTRY

DEFINITION

Conventional forcible entry is the oldest and most versatile method of entry. Usually a two-man team, using a flat head axe and the Halligan Tool accomplishes this task. It requires skill and technique to master, and at times this may have to be done by one man. When forcible entry is required, it should be started immediately. A door should be forced in such a manner as to preserve its integrity. If speed is an important consideration in gaining entry, then conventional forcible entry should be considered. Once a firefighter has mastered the skill of using the axe and Halligan Tool (Irons), most doors, even those that are well secured can be forced quickly. With the combination of the axe and Halligan, the forcible entry team can generally force any door or occupancy. It is a simple matter of technique and leverage.

ENTRY SIZE-UP
The fire ground is a very stressful place to work in. This is especially true for the first arriving units who have to accomplish a variety of tasks immediately. Among them is making a correct entry size-up.

Prior to forcing a door: The Forcible Entry Team should: **TRY THE DOOR** to determine "**IS THE DOOR LOCKED?**" Too many times over-aggressive firefighters have forced an unlocked door. They should take note of the **Type of Door and the Locking Devices** involved. Also, what are the **Prevailing Conditions** at the scene, such as heat, smoke and visibility? They should then feel the door and or the doorknob. This may give an indication of the amount of heat behind the door. Finally, **Check For Resistance**; push in at top, center and bottom of door. This may give you and idea as to where the locking devices are secured.

To master this skill a firefighter should have a basic knowledge of the types of doors and security devices he will encounter, in addition to the skills gained through hands-on experience. Also, the firefighters must have confidence in their skills that will allow them to work through any situation under pressure.

STEPS FOR FORCING A DOOR

Most conventional forcible entry involves several moves in order to accomplish the goal. In order to make it understandable, we have broken down the operation into three separate steps. Each step may have additional maneuvers, but once one understands the basic principles it is easy to follow and move quickly through the steps.

The recommended steps for forcing a door are: **GAP – SET – FORCE**.

FORCING *INWARD* OPENING DOORS (door swings away from you)

1. **GAP the DOOR -** This step will make an opening in the door and/or frame to create **a purchase point. It may also force open a poorly secured door.**
   - Work the **ADZ** into the stop on the doorframe approximately 6 inches above or below the lock (see "Note" below). The tool can be set into the frame by swinging like a bat and driving the **ADZ** into the frame.
   - If there are 2 locks close together, go between them (unless they are stacked locks).
   - Push up or down on the Halligan Tool causing the **ADZ** to rotate and crease the door. Best purchase is gained when the **ADZ** end is used on the door, not the pike.

Note: The reason for the 6-inch rule is to avoid the Halligan Tool from striking the lock. The fork of the Halligan Tool is approximately 3-inches wide and most lock bodies are also 3-inches wide.

Technique Tip: You will lose power when pushing down if the pike hits the door. You will increase spread by moving the tool up.

2. **SET THE TOOL** - This step requires the most skill. This involves working the **FORK** of the Halligan Tool into the **Gap** to spread the door away from the frame. The Halligan Tool is considered "**Set**" when the **FORK** is "**locked in**" to the inside of the doorframe.

- Position the Halligan Tool **FORK approximately six inches above** or below the lock cylinder. If the tool is too close, the **FORK** may hit the lock and will not go through to "**lock in**." If it is too far away, the door may flex and the lock will not fail.

- Place the **FORK** of the Halligan Tool, (**Bevel to the Door**) and angle the Halligan Tool to work around the doorstop. This is considered the ideal position since it produces the most spread of the door and frame and puts the most stress on the locking device. It is important for the member holding the Halligan Tool to "**walk the tool**" around the doorstop and frame.

- This method gives a greater range of motion to the Halligan Tool since the **adz** will be facing **away** from the door and not strike the door when the door is forced.

- It also offers a better striking position. The Halligan Tool will stand out at approximately 90 degrees to the door allowing the member with the axe more room to maneuver and deliver the necessary blows.

 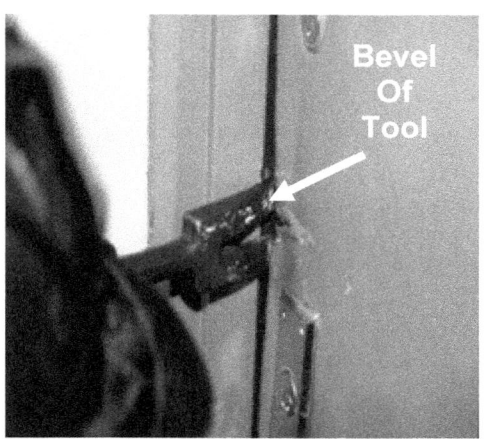 

Note: When there are multiple locks closely spaced on the door (stacked locks), position the tool above the upper lock or below the lower lock. Remember the six-inch rule is a general rule and should allow the FORK to clear the inside to the lock.

- The **forcible entry firefighter should be between the door and the tool.** Generally the forcible entry member should have his shoulder in contact with the door. This position gives a good view of the area where the tool is being driven in and also gives full range of motion for the tool as it is pushed away from the door as it is being driven in.

- The forcible entry firefighter should keep his **eyes on the FORK end** of the Halligan Tool where it is being driven into the **Gap**.

- **Keep moving the Halligan Tool away** from the door as it is being driven in (struck).

SET THE TOOL

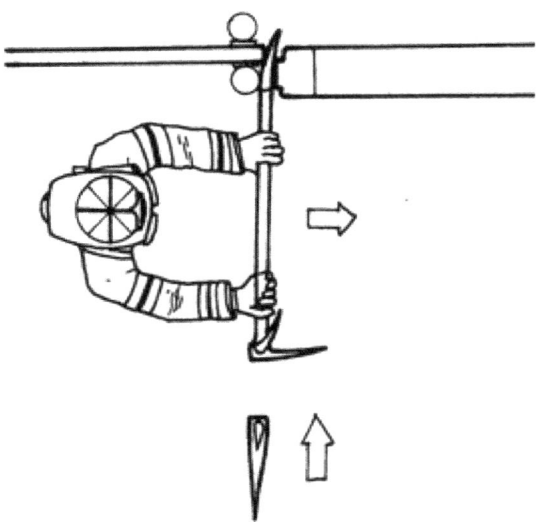 

Technique Tip: As soon as the tip of the fork is past the edge of the door, sharply push the tool away from the door. "Spring" the door away from the frame and maintain pressure on the tool to prevent the tips from striking the frame.

- **When** the Halligan is nearly **perpendicular** to the door, **drive in forcefully.** The **FORK** end of the tool is driven past the inside of the frame. This will insure the tool being "**locked**" into position and not slipping when pressure is applied.

- The tool is **SET** when the **ARCH of the FORK is even with the inside edge of the door / doorstop.**

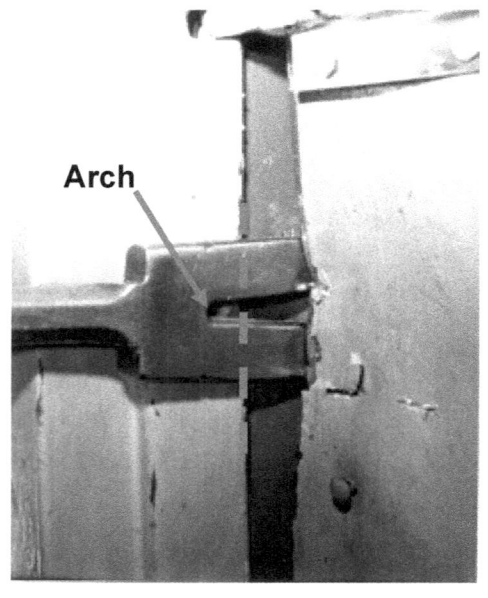 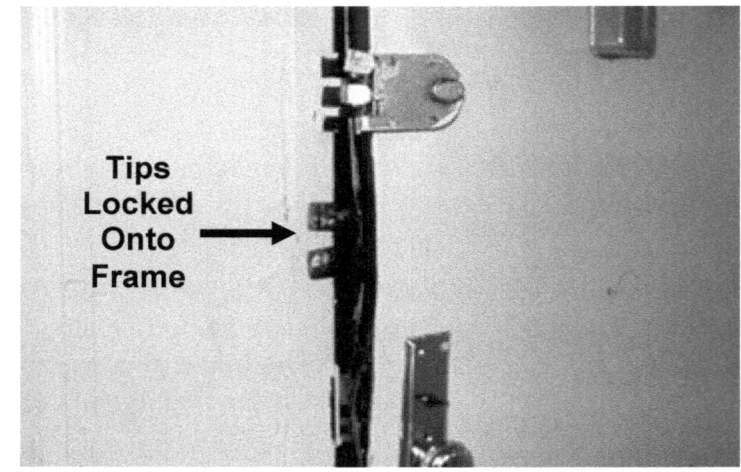

STRIKING THE HALLIGAN TOOL - Coordination and communication must be maintained between the members of the forcible entry team.
- The member holding the Halligan Tool (forcible entry firefighter) controls the operation.
- The member with the axe strikes the Halligan Tool **PERPENDICULAR TO THE ADZ.**
- The member with the axe may have to stand, crouch or kneel to obtain the best position.
- The member with the axe strikes the Halligan only when told.
- The commands "**HIT**" and "**STOP**" must be understood.

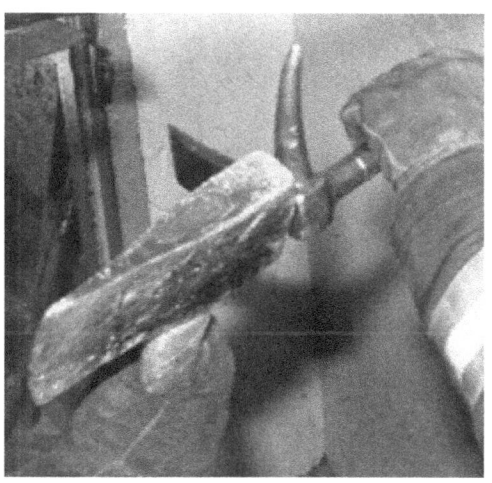

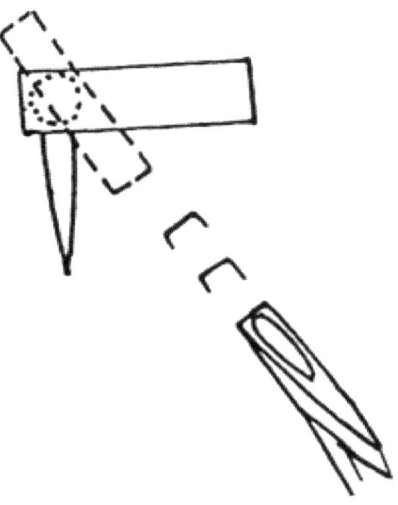

TO MAINTAIN CONTROL

- Short chopping blows.
- Perpendicular to the adz.
- In line with the shaft.

Note: As the tool is SET, more powerful blows can be delivered.

3. **FORCE** - When the Halligan Tool is set, **force** is applied to the tool creating leverage against the door.
   - Forcible entry member **changes position to Face the Door.** This gives him better position to apply pressure.
   - Ensures everyone is **ready.**
   - The other member of the team should try to control the sudden opening of the door by holding onto the doorknob or applying a hose strap to the knob.

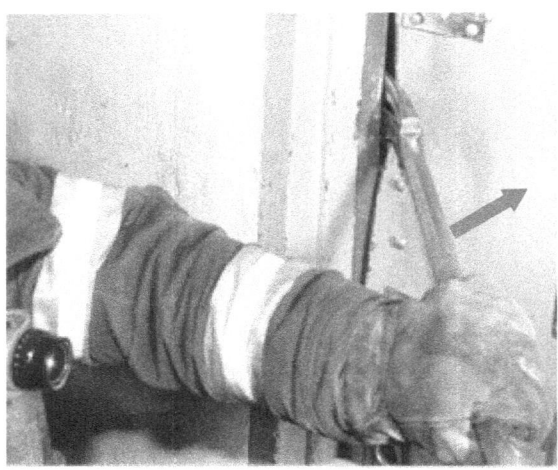
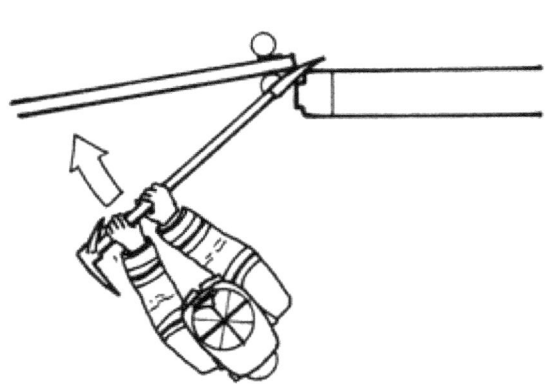

   - **Push in sharply to create maximum force.**
   - If strong **resistance is met**, a second firefighter may be used to **assist.**
   - As the door opens, the second firefighter must **MAINTAIN CONTROL OF THE DOOR.**

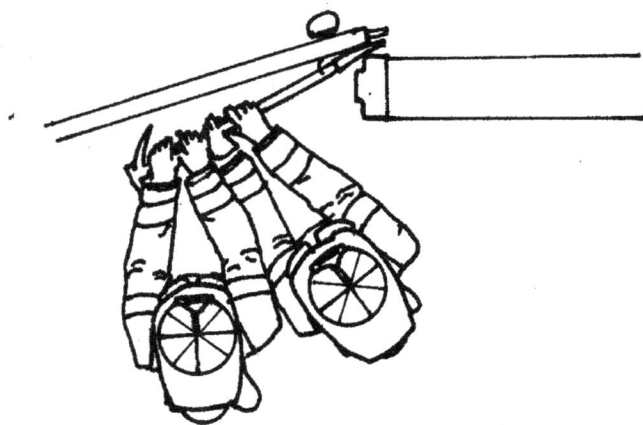

Note: In the above method, as the door is flexed from the pressure, note the presence of fire behind the door. If fire is present, make sure there is a charged line in position to protect the forcible entry team.

ALTERNATE METHODS TO GAP AN *INWARD* OPENING DOOR

Pike or Adz into the Frame

Driving the **PIKE or ADZ into the doorframe** with either the axe or maul, or simply by taking a "baseball-bat swing," should give the tool enough bite to ensure a purchase. Try to bury the **PIKE** into the frame as close to the door and lock as possible. This procedure is very quick and simple for a one-man operation. **This procedure may force the door.** It works best on wooden doors with wooden frames.

- Place the **PIKE** between the door and the doorstop, on or near the lock.
- Drive (set) the **PIKE** with the axe.
- **Push down or Pull into** the door (Gap) with the Halligan Tool.

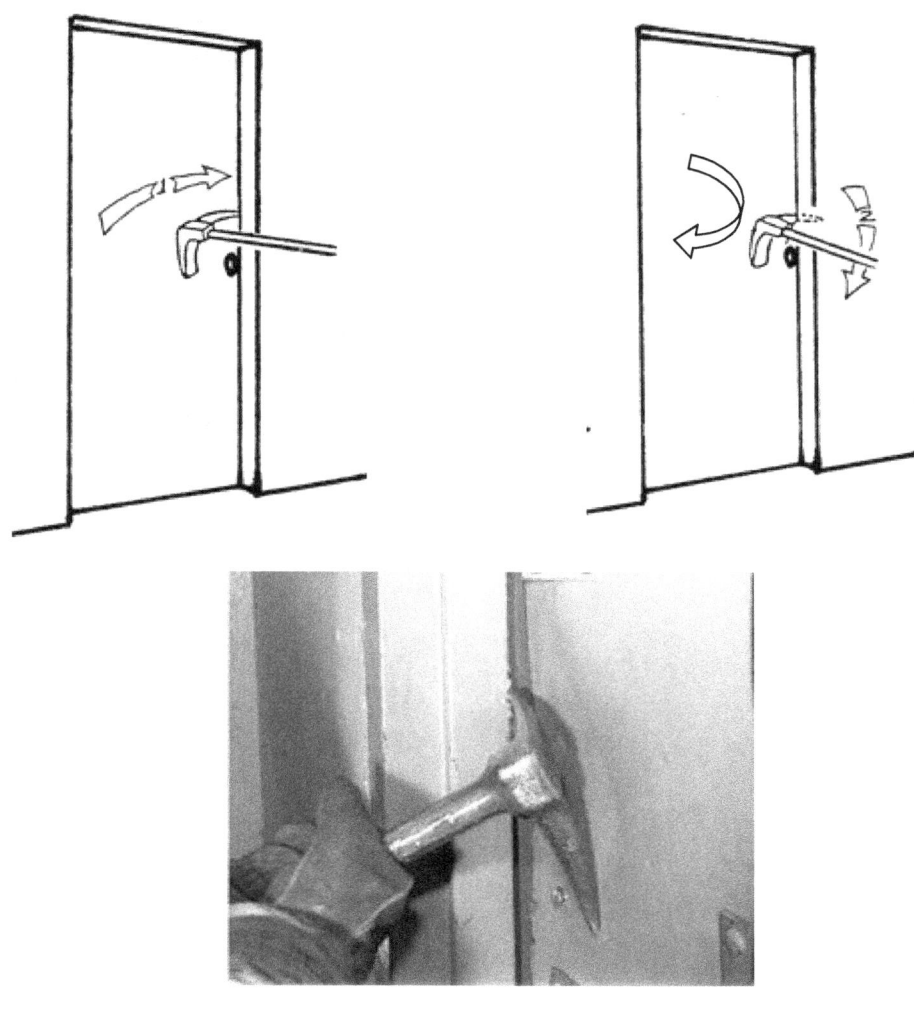

Gapping The Door

ALTERNATE METHODS TO GAP AN *INWARD* OPENING DOOR

Bevel to the Frame

- Place the **BEVEL** of the Halligan Tool **against the frame** and with an axe or maul, drive the Halligan Tool in.

  This is usually done when there is a very tight door with stiff resistance:
    - Usually a metal door with a metal frame.
    - Obstruction is in the way making it difficult to strike the tool.

  As the tool is driven in, it must not be driven **into the frame**. This takes a "**fee**l" of the tool to do correctly.

Note: **This method does not give the full range of motion to the tool. The ADZ will strike the face of the door as the member pushes towards the door.**

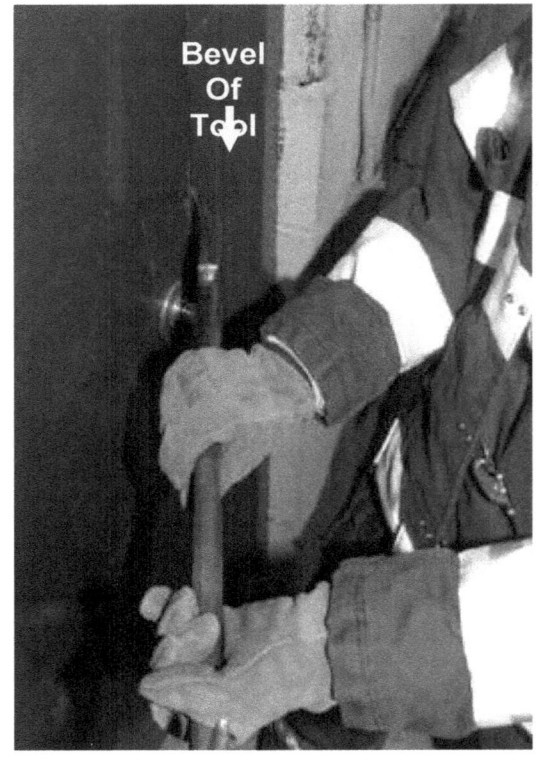

Batter the Door

Batter the Door with a few sharp blows with the Halligan Tool, axe or maul to loosen the door to allow the adz to be slipped in. However, when using this method, you must hit the **"rail of the door,"** since this is usually the strongest area of the door.

Striking the door at other areas may weaken the door or knock out a panel such as on a raised panel door. **This is dangerous** since it allows heat, smoke and fire to vent out of the opening making further forcible entry more difficult. **Do not** knock in the panel unless there is a charged line in position.

Batter the Door

Technique Tip:
> If the door is set in a weak wood frame, several **sharp blows to the door right on the lock** may split the frame. This is especially true if the door contains a **mortise lock**. Note the mortise lock is set into a cavity made in the door. This may compromise the integrity of the door.

Batter the Door Frame

Batter the Door Frame by striking with an axe, maul or Halligan Tool approximately 6 inches above or below the lock and driving it away from the door to allow entry for the Halligan Tool. Some times steel frames are filled with concrete and may not crush.

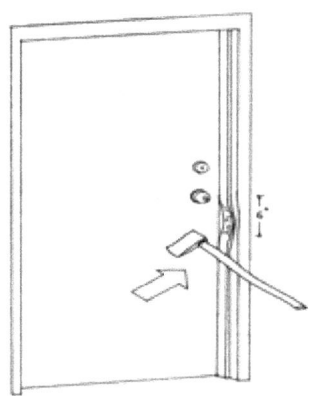

Remove the Door Stop

Remove the doorstop on wood and / or Kalameine doors with the **ADZ** or **FORK** end of the Halligan Tool. This is a simple way to open a door with minimal damage. This method works best on wood doors with wood frames.

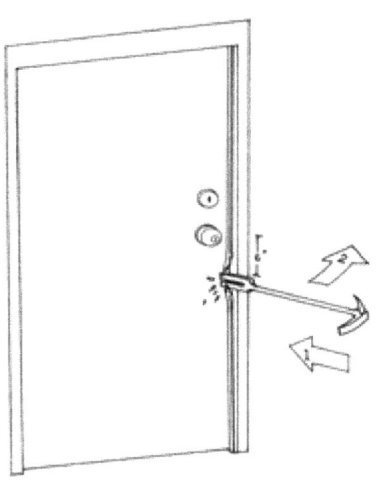

THE HALLIGAN TOOL GETS STUCK

Problem: THE FORK IS IN CONTACT WITH THE DOORFRAME

Solution:
- Increase the angle away from the door.
- "Rock" the tool to free it.
- Re-Gap the door; reverse the tool (Bevel to Frame).
- Move further away from the lock; this makes the door easier to spread.

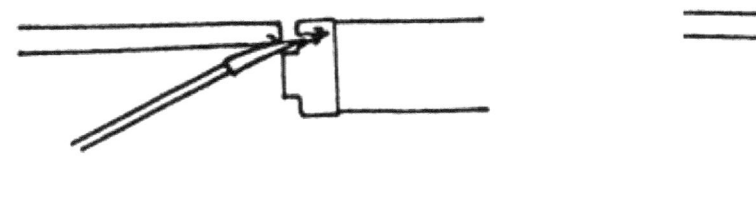

Contact With Doorframe   **Increase Angle**

Problem: THE FORK IS HITTING THE BOLT OR LOCK

Solution:
- Reposition the Halligan Tool above or below the lock.

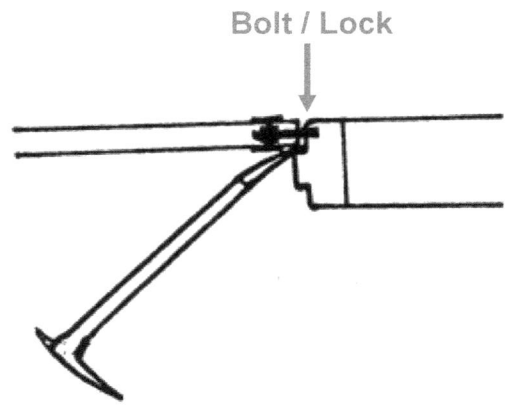
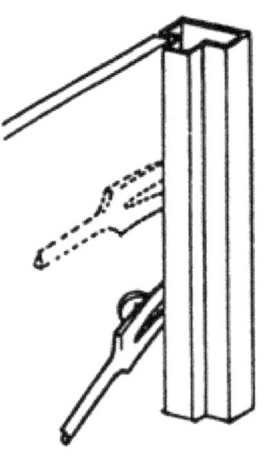

Striking Bolt / Lock   **Reposition Halligan**

THE HALLIGAN TOOL GETS STUCK

Problem: THE FORK IS WEDGED INTO A TIGHT DOOR

Solution:
- **Springing the Door**
    - Move the Halligan Tool **side to side** to free up the tool.
    - Push sharply and hold until the tool is driven further in.

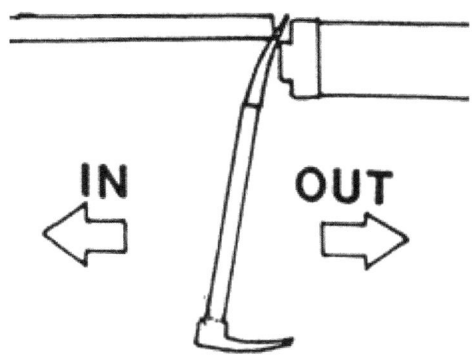

Springing The Door

Problem: THE FORK IS WEDGED INTO A TIGHT DOOR

Solution:
- **Slipping The Lock**
    - Move the Halligan Tool **up and down.** This may allow the tool to slip past the bolt of the lock.

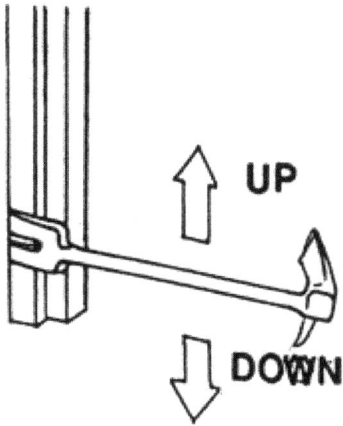

THE DOOR DOES NOT OPEN DURING THE INITIAL OPERATION

Problem: **THE DOOR FLEXES AND DOES NOT OPEN**

Solution: Method A – Using The Adz

- **Maintain the purchase** with axe or other tool.
- **Slip the ADZ (or door chock) inside and behind the doorframe.**
- Both members of forcible entry team **push in or pull** on the Halligan Tool.
- If the doorframe collapses and the adz gets stuck between the door and the frame, use the axe to wedge open the space, then **push or pull** the Halligan away from the door to release the adz.

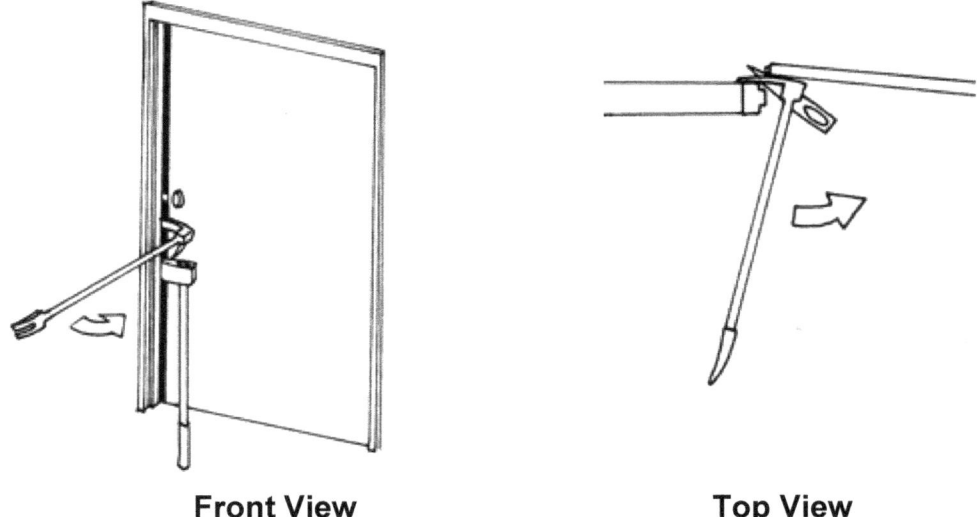

Front View   **Top View**

Note: This method greatly increases the Range of Motion of the Halligan Tool and will break most locks.

Problem: **THE DOOR FLEXES AND DOES NOT OPEN**

Solution: Method B – Using The Axe and Fork End

- Extra push may be obtained by placing the **head or blade of the axe between the Halligan and the door.**
- Place either the blade or the head of the axe into the door seam.
- Push in sharply with the Halligan.

THE DOOR DOES NOT OPEN DURING THE INITIAL OPERATION

Method B – Using the Axe and Fork End

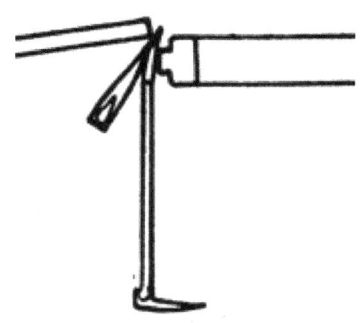

Blade Into Door Seam

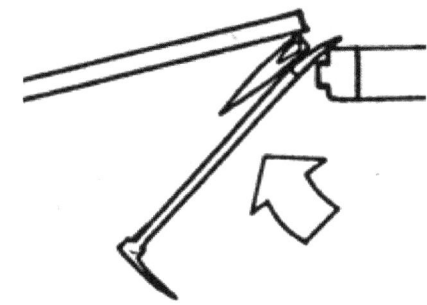

Head Into Door Seam

DRIVING THE LOCK OFF THE DOOR

Problem: **DOOR OPENS PARTIALLY DUE TO STRONG LOCK(S)**

Solution:
- Place the Halligan Tool **directly on the lock and drive it off** the door. Driving the lock off the inside of the door takes sharp blows with the axe. Remember that you are trying to drive out the screws that hold the lock onto the door.

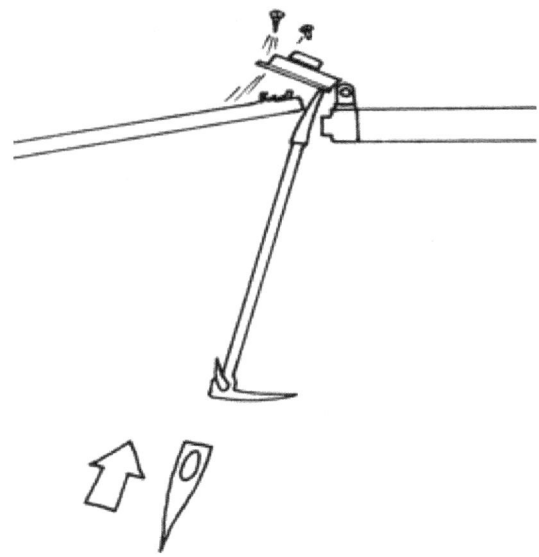

DRIVING THE LOCK OFF THE DOOR

 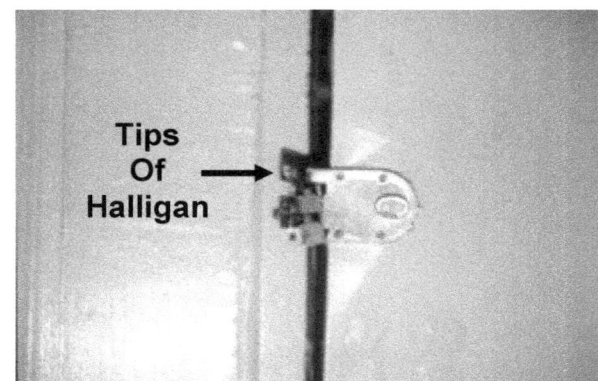

Note: If you can crush the door enough to see the locking device (especially the vertical deadbolt type), you may be able to shear off the striker with the Halligan Tool.

ANGLE IRON *INWARD OPENING DOOR* (Door swings away from you)

- Usually bolted to the door, may be partial or full length.
- The angle iron may be flat stock or shielded (interlocked with "J" channel).
- Place the **BEVEL** towards the angle iron and the tool **PERPENDICULAR** to the door between the angle iron (shield) and the frame.
- Lock the tips of the fork into the doorstop and push in sharply, **(GAP) the door with the fork between the angle iron and the frame.**
- Reset the tool and drive in (**SET**).
- Using the angle iron under the tool, (**FORCE**).

  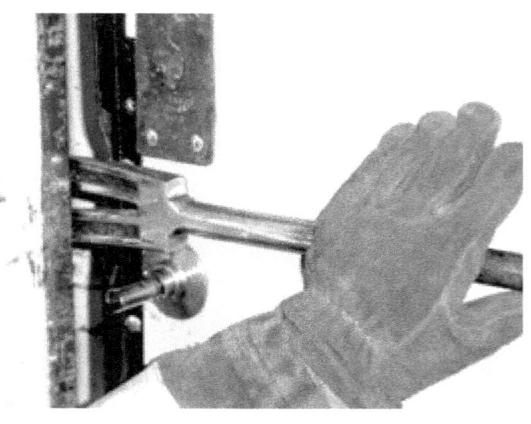

FORCING THE "J CHANNEL" *INWARD* OPENING DOOR

- A newer type of device that is screwed into the door frame.
- The technique is modified by driving the **FORK** end of the Halligan between the shield and the door frame.
- Drive the **FORK** in until the tips hit the door.
- Push the tool toward the door, popping the shield off the frame or bending it out of the way.
- Re-set the tool and drive it in until it is set.
- Force the door.

NARROW HALLWAY OR RECESSESED DOOR

When there is little room and/or limited visibility to swing an axe, an alternate method of striking the Halligan would be:

- Slide the head of the Axe **along the shaft** of the Halligan Tool and **strike the shoulder** of the fork end.

- **Bevel to the Frame** may give better results due to easier entry and better angle to strike the tool.

 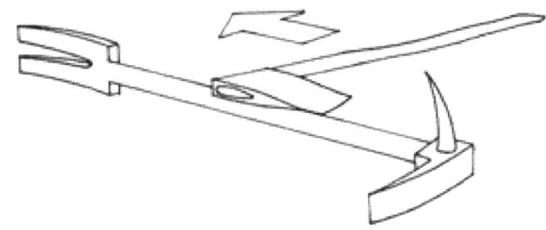

Note: The shoulder of the fork end of the Halligan must be squared off to do this. See Chapter 16.

FORCING *OUTWARD* OPENING DOORS (Door swings toward you)

Using The Adz End

- Place the **ADZ** between the door and the frame.
- **GAP** the door by rocking the tool **up and down** to spread the door from the frame.
- **SET** the tool, and pry the door out by **pulling** on the Halligan so the **ADZ** can be driven in. Be careful not to **"bury the tool"** into the doorstop. See Chapter 16.
- **Force** the door, set the **ADZ** end around the inside of the door.

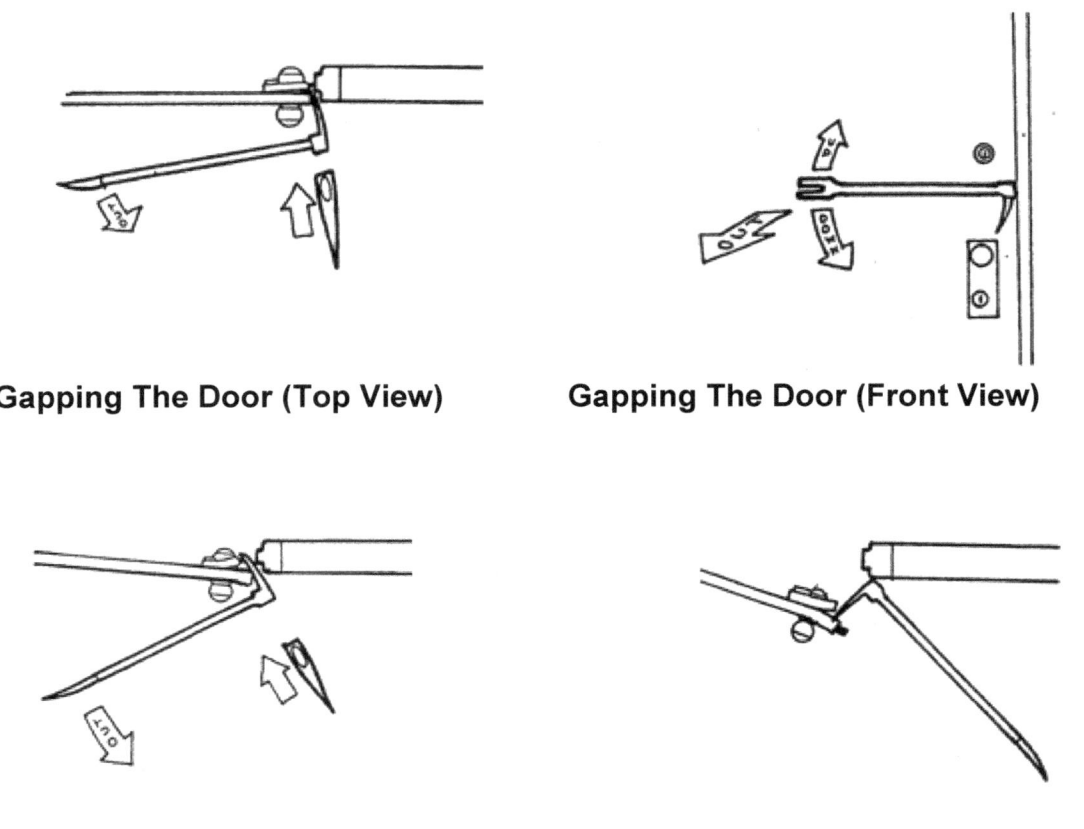

Gapping The Door (Top View)   **Gapping The Door (Front View)**

Set The Tool (Top View)   **Force The Door (Top View)**

Note: The firefighter faces the door at all times.

FORCING OUTWARD OPENING DOORS (Door swings toward you)

Using the Fork End

- **GAP** the door by placing the bevel side of the **FORK** toward the frame, just above or below the lock or hinge.
- **SET** the tool, pry the door by **pulling** out on the Halligan so the **FORK** can be driven in past the inside frame. Be careful not to **"bury the tool"** into the doorstop. See Chapter 16.
- **FORCE** the door, set the **FORK** end around the inside of the door and by **pulling or pushing** the Halligan Tool *away* from the door (toward the wall).
- In order to use this method, the Halligan Tool must have sufficient room to allow the movement of the tool away from the door.

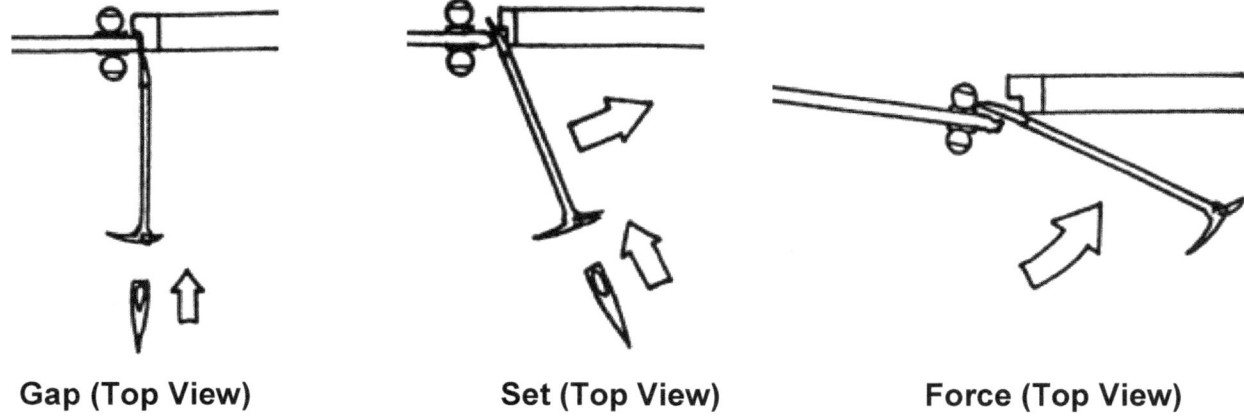

Gap (Top View)   **Set (Top View)**   **Force (Top View)**

Note: These methods will be dictated by the configuration of the building or any obstructions near the door.

PROBLEMS ENCOUNTERED WHEN FORCING *OUTWARD* OPENING DOORS

Problem: RECESSED DOOR OR OBSTRUCTION

Solution:
- To allow the **ADZ** to be driven in and around the door stop and to provide sufficient space for the **ADZ** to move away from the door.
  - Make a hole in the wall (if possible), for the movement of the tool.
  - **GAP – SET – FORCE** the door.

PROBLEMS ENCOUNTERED WHEN FORCING *OUTWARD* OPENING DOORS

Problem: **DIFFICULTY GETTING A PURCHASE**
(Tight Seam between Door and Frame)

Solution:
- Use the **Blade** of the axe.
- Use the **Fork** or **Adz** end of the Halligan.
  - Tilting the **Adz** up or down may start the purchase easier.

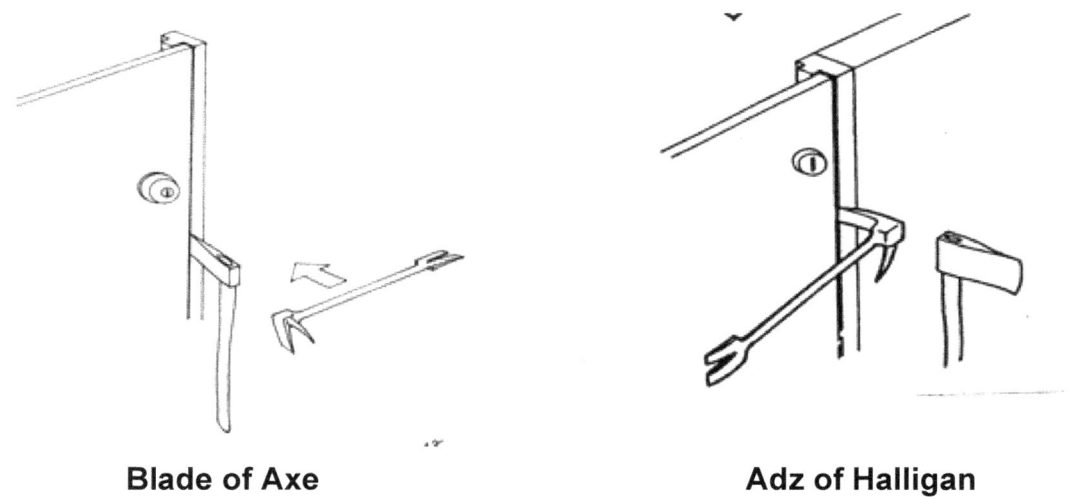

Blade of Axe   **Adz of Halligan**

METAL STRIP ON THE EDGE OF THE *OUTWARD* OPENING DOOR

Additional security may be installed on these doors by bolting a metal shield to protect the space between the door and the frame. It may be a full-length or partial shield. Dealing with the shield will require an additional step before proceeding to **Gap – Set – Force.**

METAL STRIP ON THE EDGE OF THE *OUTWARD* OPENING DOOR

- Drive the **ADZ** end under the edge of the metal strip and push the tool toward the door. Work the **ADZ** between the door and the frame, and drive in to establish a gap.

- Drive the **FORK** end under the edge of the metal strip and push the tool back toward the door.
- Work the **FORK** between the door and frame. Reverse the tool if necessary.

Reverse the Tool if Necessary

Brute Force Entry

METAL STRIP ON THE EDGE OF THE *OUTWARD* OPENING DOOR

- Drive the **ADZ** end between the door and shield, bending the shield away to allow entry of the Halligan Tool.

- Shear the bolts and pry, bend or remove the shield as a last resort.

Note: At times if the angle iron is secured well, it may assist you in opening the door. If not, then you have to remove it to access the door.

HYDRAULIC FOCIBLE ENTRY TOOLS

HYDRAULIC TOOLS

These tools are designed for **doors that open inward** (away from you), and have also been used successfully on **sliding elevator doors**. They work best on doors with **strong metal frames**.

These tools are quite easy to use. The two most popular models are quite different in design. One type, The Rabbit Tool, is a two piece unit, consisting of a spreader and a pump. They are connected with a short length of high-pressure hose. The second type is a one-piece unit that incorporates the pump and spreader together. Equally effective they have found a solid footing in the fire service.

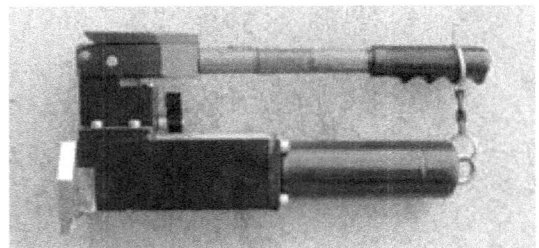
Hydra-Ram

Rabbit Tool

RECOMMENDED STEPS FOR FORCING A DOOR

- **GAP THE DOOR** – Using the **ADZ** end.

- **SET THE TOOL** - Insert the jaws between the door and the frame midway between the knob and the lock; the **jaws must be in the closed position.**

- **FORCE THE DOOR** - The door should open with several pumps of the handle.

Gap

Set

Force

Note: **When there are multiple locks, insert the jaws between the knob and lock and then move to the proximity of the next lock.**

THE DOOR DOES NOT OPEN DURING THE INITIAL OPERATION

Problem: **TOOL IS FULLY EXTENDED AND THE DOOR STILL DOES NOT OPEN**

Solution: **Reposition Tool On The Lock**
- Locate the locking device.
- Wedge the door open with the head of the axe or the Halligan Tool.
- **Reposition the hydraulic tool** directly on the lock and extend.

Solution: **Drive Lock Off The Door**
- Maintain the opening with the hydraulic tool.
- Drive the lock off the door with the axe and Halligan Tool.

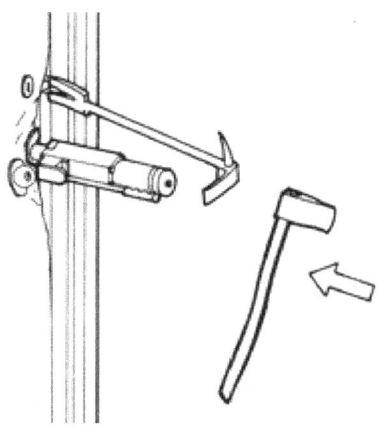

Drive Lock Off

THE DOOR DOES NOT OPEN DURING THE INITIAL OPERATION

Problem: **TOOL IS FULLY EXTENDED AND THE DOOR STILL DOES NOT OPEN**

Solution: **Reposition Tool Inside The Doorframe**
- Maintain the opening with the hydraulic tool.
- Slip the axe or Halligan Tool into the gap and **maintain purchase.**
- Reposition the hydraulic tool **INSIDE THE DOORFRAME.**
- Pump the handle to extend the tool.

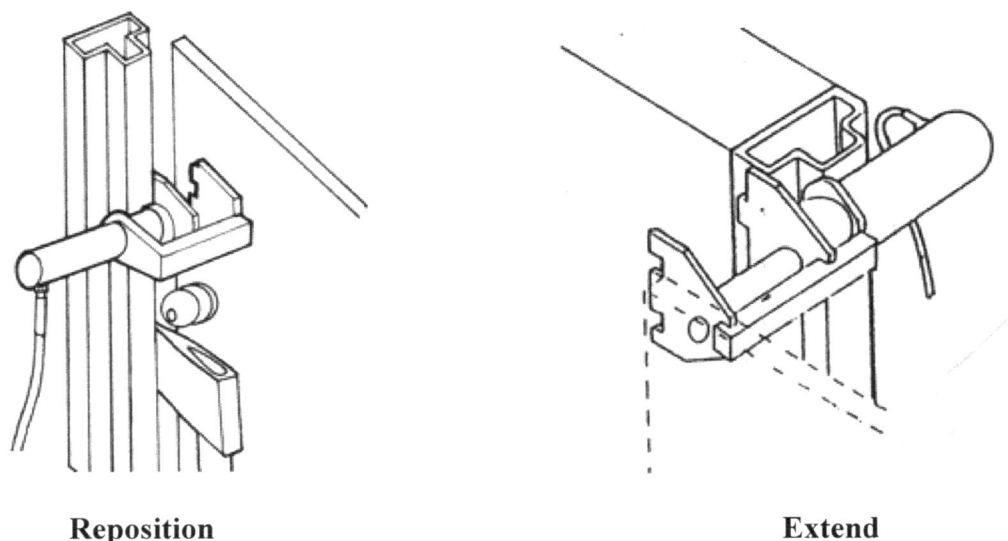

Reposition   **Extend**

Note: This maneuver may not work with the Hydra-Ram tool.

THE DOOR DOES NOT OPEN DURING THE INITIAL OPERATION

Problem: **DOOR NEARLY FORCED BUT NEEDS A LITTLE MORE**

Solution: **Use Head of the Axe to Extend the Spread**
- Maintain the purchase with the Halligan Tool.
- Place the head of the axe between the door and the jaw and extend.

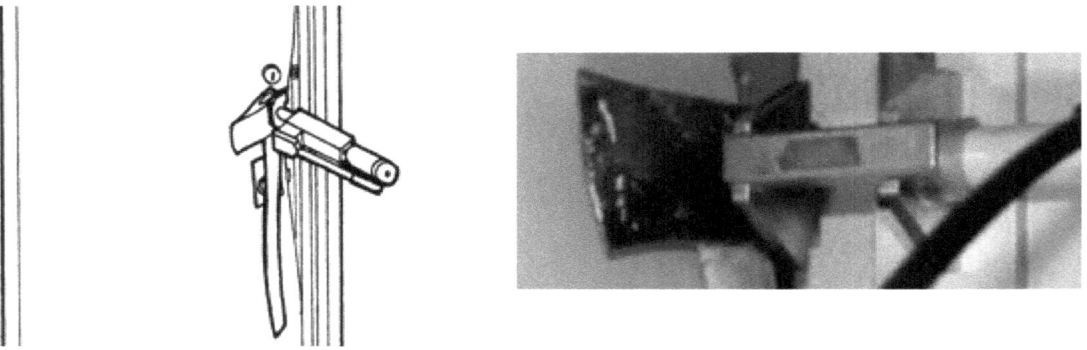

ANGLE IRON *INWARD OPENING DOOR* (Door swings away from you)

- Usually bolted to the door, may be partial or full length.
- The angle iron may be flat stock or shielded (interlocked with "J Channel").

Partial Angle Iron

- **GAP** the door by placing hydraulic tool above or below angle iron and open to full extension.
- Maintain this purchase with the axe or Halligan Tool.
- Reposition the hydraulic tool between the door stop and the edge of the angle iron (on the angle iron).

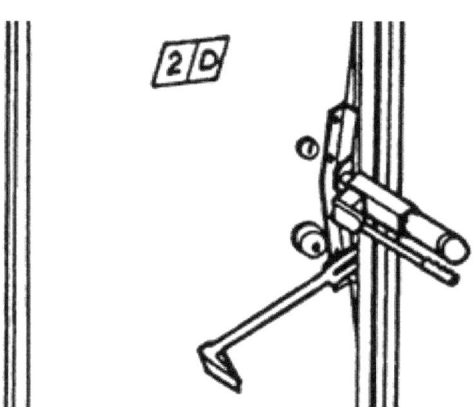

Full Length Angle Iron

- **GAP** the door by driving the **FORK** end of the Halligan Tool between the angle iron and the frame, pushing the Halligan **towards** the door.
- Reposition the hydraulic tool on the angle iron and extend.

MAGNETIC LOCK

The doors these type locks are installed on are usually *outward* opening type doors. This is an exception to the use of the hydraulic forcible entry tool, which is primarily designed for *inward* type opening doors.

Procedure for Forcing Magnetic Lock
- Place Halligan Tool through the door handle with **ADZ** end toward you.
- Place the jaw of the hydraulic tool behind the **ADZ** end of the Halligan Tool.
- Pump the tool using the doorframe as the base and **"pull"** the door away from the magnetic lock.

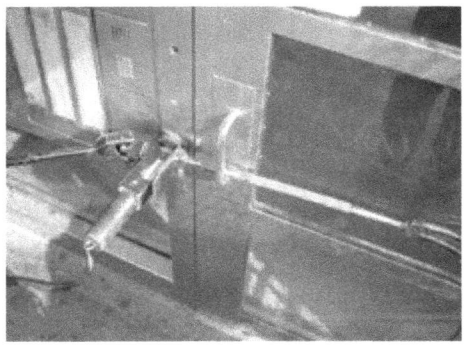 

Note: The Hydra-Ram tool has enough power to overcome the force of the magnetic lock.

MULTI-LOCK DOOR

These type doors, as described in Chapter 6, are formidable doors. Listed is the suggested order of tool placement in forcing these type doors.
- Force the lock side pin first (door knob side).
- Force the top pin second.
- Force the bottom pin third (may be forced by placing the hydraulic tool at the knob/lock side bottom corner of the door).
- If necessary force the hinge side fourth.

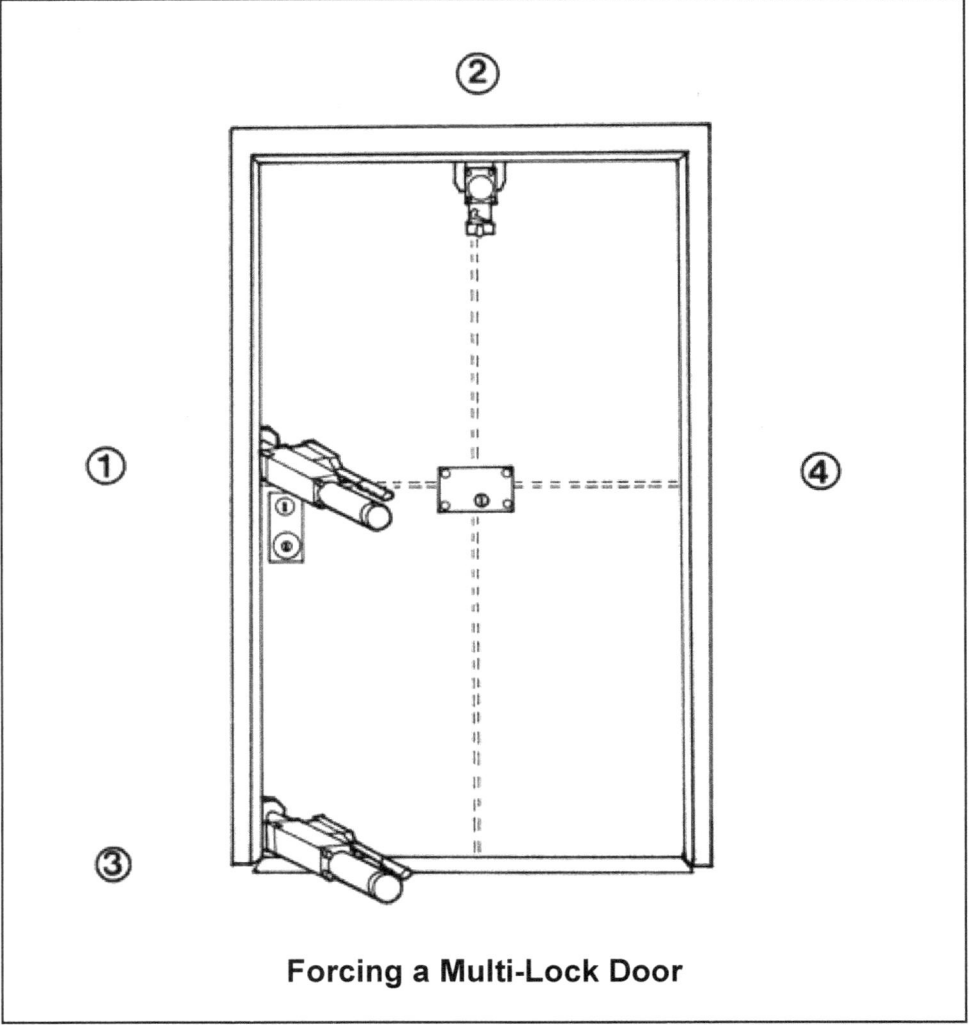

Forcing a Multi-Lock Door

Note: This may also work on the "Door Club."

HINGES

HINGES

TYPES: There are many types of hinges used today. The types we discuss will be known as:
- Standard
- Self-Closing
- Pin Type

- **STANDARD HINGES:**
    Most common type found in residential occupancies. May find two or three mounted on a door. The center pin connects the two pieces of the hinge.

- **SELF CLOSING HINGE:**
    This hinge is more common in commercial type occupancies. It is a sealed, spring-loaded hinge. These may also be mounted in sets of two or three to a door.

- **PIN TYPE HINGE:**
    As a rule these hinges are mounted on the exterior of Commercial Occupancies. The "Pin" is secured to the frame and the hinge is secured to the shutter or door.

- Forcing a door at the hinge side **SHOULD NOT** be a **primary** means of gaining entry.
- Once a door is forced in this manner you will **"lose the integrity"** of the door.
- The **PRIMARY** means of gaining entry should be on the **LOCK** side.
- Forcing a door at the hinge side should only be done when **ALL** other means of gaining entry on the lock side have failed.

STANDARD HINGE – *INWARD* OPENING DOOR (Door swings away from you)

Some suggested means of gaining entry:
- Force the door to expose the hinge, using the Halligan, then work directly on the hinge.
- Create a gap and use either the **ADZ** or **FORK** end of the Halligan.
- Place end of tool just below the hinge.
    - **ADZ** end apply force either up or down.
    - **FORK** end apply force either toward or away from the door.
- Using the **PIKE** end, as a fulcrum, separate the hinge from the frame.
- **"Batter the door"** at the hinge.
    - With the back of the axe, maul or Halligan Tool, strike the solid part of the door adjacent to the hinge.
- Hydraulic tool would be used the same as if you were forcing a door on the lock side, only place the tool on the hinge side.

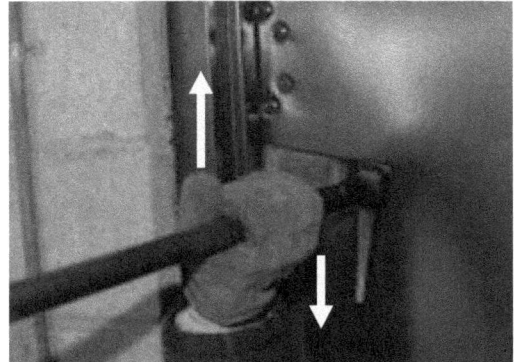
Adz End

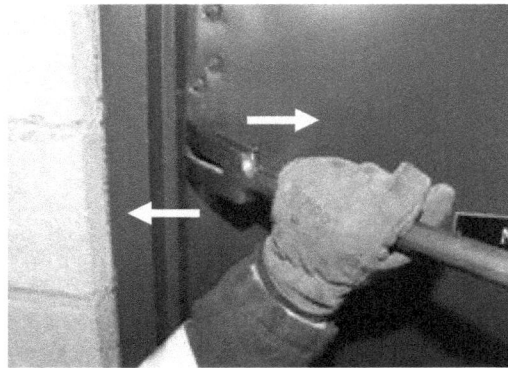
Fork End

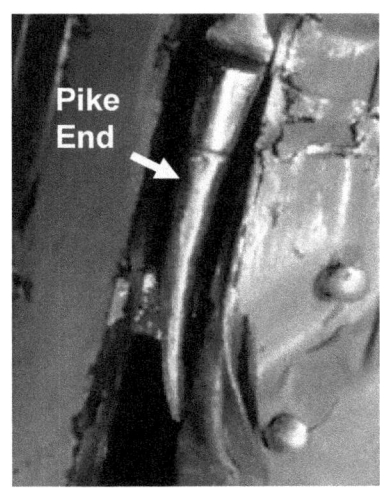
Pike End

Batter the Door

Hydraulic Tool

STANDARD HINGE

Removing A Door
- With the door partially open, slip the **ADZ** between the door and the frame just below the hinge; then pry up or down.

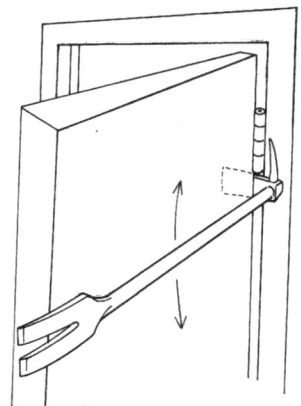

Note: **ALWAYS attack the UPPER hinge FIRST so that smoke and heat will rise while completing the entry on the bottom of the door. Be aware, many doors now have three hinges.**

STANDARD HINGE – *OUTWARD* OPENING DOOR (Door swings toward you)

Some suggested means of gaining entry:
- Place the **FORK** end of Halligan Tool over the exposed hinge and pry up or down.
- On stronger hinges drive the Halligan over the hinge and twist side to side to break or loosen the mounting screws, then pull out.
- Remove the pin if possible to separate the hinge.

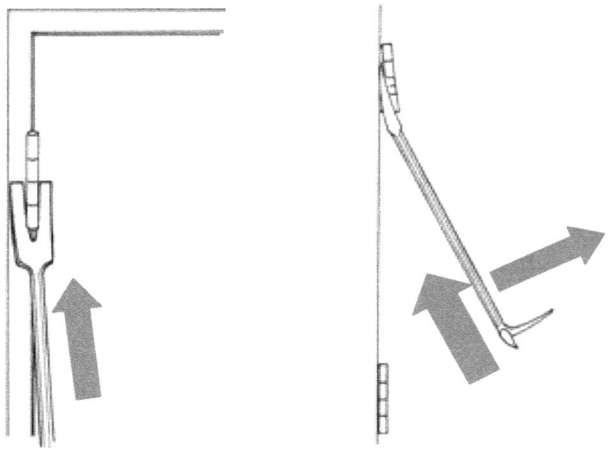

Note: **ALWAYS attack the UPPER hinge FIRST so that smoke and heat will rise while completing the entry on the bottom of the door. Be aware, many doors now have three hinges.**

Technique Tip:
   For a Bulkhead door, keep the door between you and the opening to protect from heat and or flames, which may come out.

SELF-CLOSING HINGES

These hinges may be found anywhere, but are very often found on bulkhead doors. They usually have a threaded rod with two cap nuts, which can be easily unscrewed with a pair of channel locks or a snap-on cap, which may be pried off. Some suggested ways to force these hinges are:

Method 1:
- Unscrew or snap off, the top cap and tap the threaded pin down.

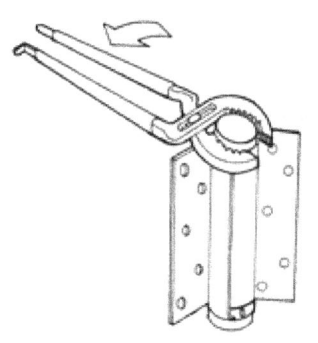 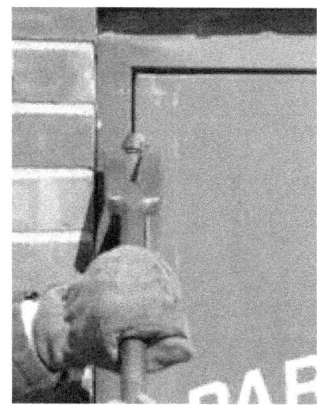 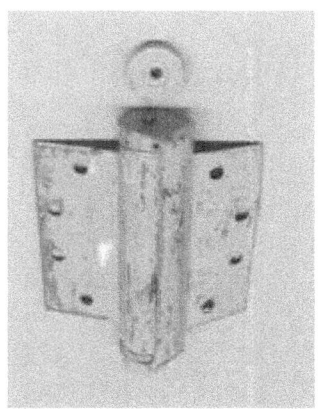

- Pull the bottom cap and pin down and out.

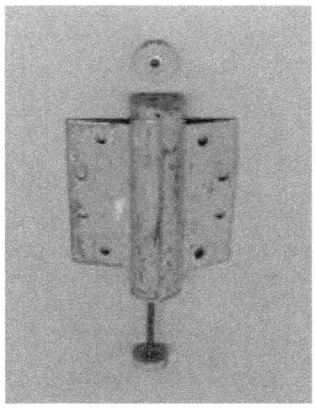 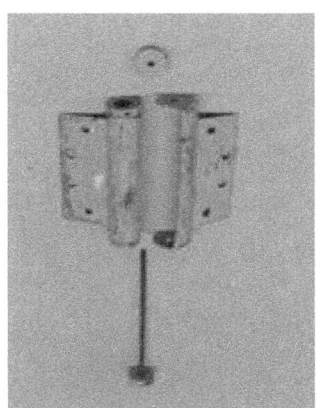

Method 2:
- Drive the **ADZ** or the **FORK** of the Halligan Tool into the body of the hinge between both sections, splitting the hinge into two pieces.
- Pull up or push down with the Halligan Tool.

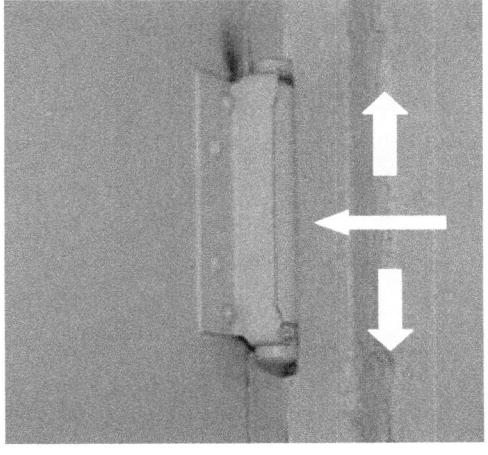

SELF CLOSING HINGES

Method 3
- Cut the hinges with the forcible entry saw.

Note: ALWAYS attack the UPPER hinge FIRST.

Technique Tip:
> For a bulkhead door, keep the door between you and the opening to protect from heat and or flames which may come out.

PIN HINGES

These types of hinges are usually found on shutters. They can also be found on commercial buildings and places of public assembly. The "pin" is attached to the window frame or doorframe, and the shutter or door holds the corresponding hinge. Some suggested means of forcing entry:
- Use the power saw to cut the hinge.
- Breaking the anchor point where the hinge is set by using the back of the axe, maul or Halligan.
- Prying the hinge with the hydraulic tool.

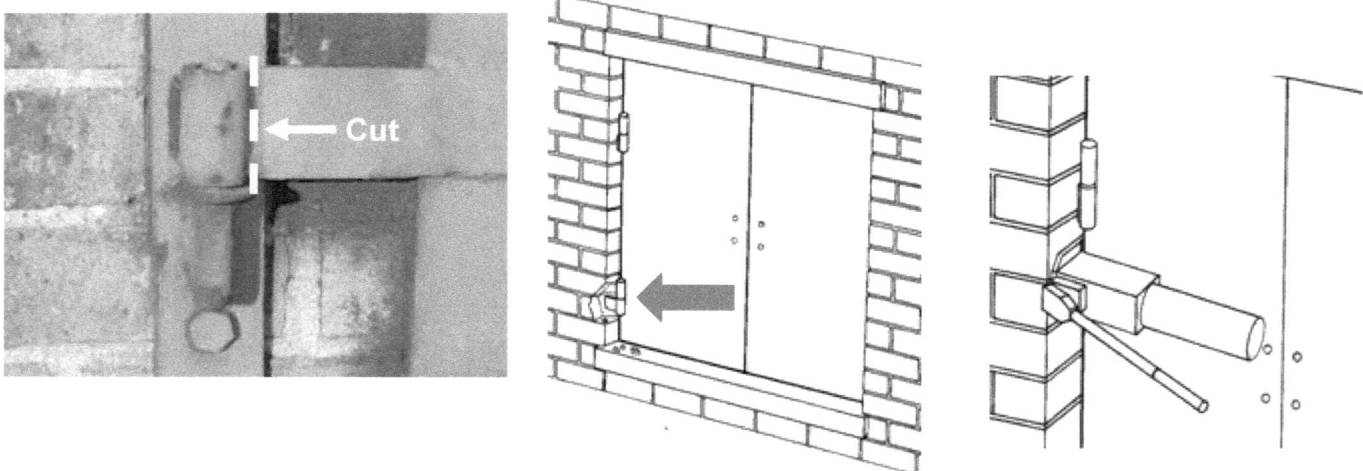

Note: Be aware of possible venting smoke and or fire. Place the shutter/door between you and the opening if possible.

CHOCKING THE DOOR

CHOCKING THE DOOR

This is a very basic and important task that gets overlooked from time to time. Many doors are self-closing and if not chocked open it delays other members from entering the fire building (occupancy).

Whatever means used to chock the door must be **"positive,"** not something that can be knocked out unintentionally. However, it must be something that **can be removed quickly** if necessary.

It is suggested that the first unit to enter the fire building be responsible for **"chocking"** the door. That could be the Officer or any member of the forcible entry team.

Some suggested methods of chocking a door:
- A wooden chock wedged under the door. Every member should carry at least two wood chocks in their pockets.

- Head of axe slid under the door. As the forcible entry team enters the occupancy, the axe is wedged under the door. This method marks the door, keeps it open and safeguards the axe since it is rarely used **INSIDE** once the door is forced. If you feel the axe might be needed **INSIDE** then this method would not be appropriate.

- Head of axe placed between the door and the frame, below the bottom hinge. A variation of the above method is placing the axe between the hinges. This insures the door staying open and lessens the chance of the axe being kicked out by members entering. It also marks the door and safeguards the axe. If you feel the axe might be needed **INSIDE** then this method would not be appropriate.

CHOCKING THE DOOR

- Placing the hydraulic tool against the open door. Once the door is forced, the hydraulic tool is not needed and can be used to maintain the door open.

- A hook chock placed over the hinge.

- A nail placed between the frame and door.

CHOCKING THE DOOR

- Disable the door by placing a tool between the frame and the door just above or below the hinge and prying down will break most hinges and keep the door open.

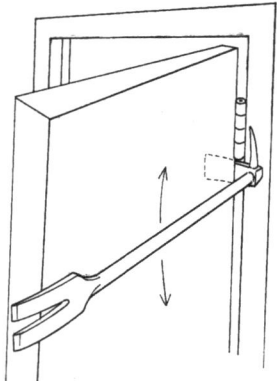

- Any other method that keeps the door open.

SECURING THE DOOR

To insure that the "opened" door does not close and re-lock, the following methods may help.

- **"Rigging the Lock"** - Throw the bolt of the dead bolt lock so the extended bolt will prevent the spring latch from locking.
- Placing a latch-strap over the door knobs to prevent the spring latch from engaging.

Rigging the Lock

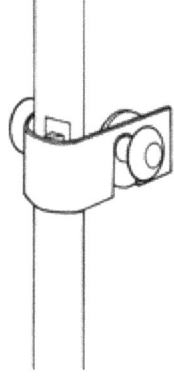
Latch Strap

THRU-THE-LOCK ENTRY

THRU-THE-LOCK ENTRY

The "Thru-the-Lock" approach is a means of gaining entry by attacking the locking device and **opening the door with little or no damage to the door or frame.** This is a professional method of entry and serves as a good public relations tool.

In most cases, this method would only be used when **time and fire conditions are not urgent,** or where conventional methods would cause more damage than the fire itself. Examples would be high rise office buildings, hotels, motels and/or commercial occupancies, where many rooms and or occupancies must be checked without causing too much damage. The Thru-the-Lock method usually does not create as much of a security problem as conventional forcible entry methods.

There are times with certain types of locks that the Thru-the-Lock method of forcible entry may be a quicker, more efficient means of entry, whatever the conditions.

Since security has been improved through technology over the years, this book can only address what is most common. This chapter of the book will outline some basic principals, methods and techniques used in **Thru-the-Lock Entry.**

Size-Up

Size-up is an important function that is performed, for all operations, on the fireground. It is critical that a proper size-up is done before we begin our forcible entry operation.

Though it is impossible to know for sure what type of lock is securing the occupancy by looking at a solid door from the outside, we can make an educated guess based on:
- Type of **occupancy.**
- Type of **door.**
- **Location** of the **lock cylinder(s).**
- **Direction** the door **moves** (inward or outward).
- What we **see on the door** (other than the locks).
- Anything **unusual** (lock cylinders out of line).
- Knowledge of the **type of lock.**
- Let the **fire condition** dictate your method of entry.

Combine all of this information with past experience and proceed in **attacking the lock,** not the door.

We need to understand that only **practice** will make us more proficient in our operation, so we must use every opportunity.

Note: The cheaper the lock, the more difficult it may be to force. Cheaper locks have a tendency to break up causing delays, and/or requiring alternative means of pulling the cylinder.

THRU-THE-LOCK METHOD

KEY-IN-THE-KNOB LOCK

As the name implies, the locking mechanism is part of the knob. These devices are found on both inward and outward swinging doors. The spring latch on the majority of these locks enters the striker approximately 1/2 inch.

FORCING THE KEY-IN-THE-KNOB LOCK - Using the Officer's Tool

The doorknob can be removed simply and quickly with the Officer's Tool, without damaging the stem assembly.
- If the **door is hollow,** an axe can be placed behind the tool to give the fulcrum a substantial base to pivot off.

After the doorknob is removed, insert the stem of the Key Tool into the slot (if present) or into the back of the spring latch and pull or twist toward the hinge side of the door to activate the latch.

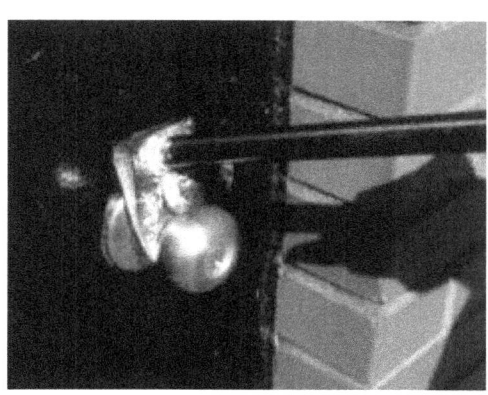

FORCING THE KEY-IN-THE-KNOB LOCK – Removing the Center of the Knob

There are some locks where the center of the knob can be removed (example, Kwikset type lock) with a knife-like tool or slotted screw driver.

By using the **"Bam-Bam Tool"** you can pull the face of the lock to expose the stem slot inside the knob where the correct Key Tool can be inserted.

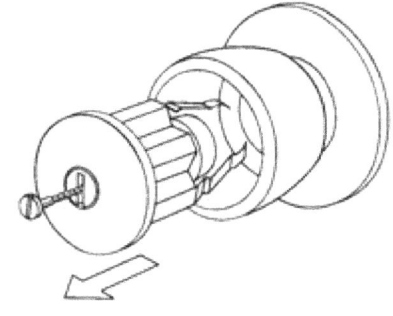

FORCING KEY-IN-THE-KNOB LOCKS – Outward Swinging Doors

Key-in-the-Knob locks on outward swinging doors have a simple spring latch which can be slipped back (opened) with a flat tool such as a **ShoveTool.**

At times there is a simple device known as **anti-loitering pin**, which may be added to the latch. This pin prevents the insertion of the shove tool without moving this pin first.

Pin Engaged   **Pin Pushed Back**

TUBULAR DEAD BOLT

This is a very popular locking device. It may be single or double key activated. It is a cross between a mortise lock, rim lock and a key-in-the-knob lock. These locks may be recognized by their **position on the door and/or the size and shape of the cylinder.**

TUBULAR DEAD BOLT

These locks have become more sophisticated as the demand for greater security has increased. They may have a hardened steel rod through the center of the locking bolt. The length of the bolt has been increased to the point that it **may take two full rotations of the key to remove the bolt from the keeper**.

The lock face is usually held in place by a hardened steel mounting. The cylinder is either **too deep or too wide**, which prevents the K-Tool from being used. In order to use the Thru-the-Lock method, the cylinder needs to be removed to enable the use of a Key Tool to trip the lock. If the K-Tool is unable to remove the cylinder, then an alternate method of removal would be needed in order to use this method.

If the cylinder is unable to be removed then you will have to resort to conventional forcible entry methods to force the lock.

The stem of the tubular deadbolt, which retracts the locking bolt, can be various shapes.

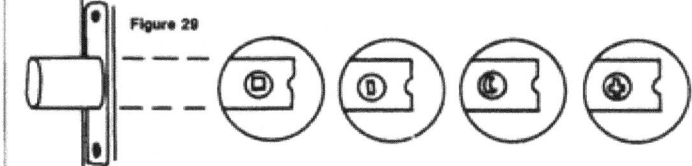
Figure 29

FORCING THE TUBULAR DEAD BOLT

- **Remov**e the cylinder by pulling it out with either the Officer's Tool, K-Tool or modified Halligan.
- Insert Key Tool.
- Rotate to open.

Technique Tip: Place the Officer's Tool at an angle to start the operation.

Note: Using the Officer's Tool would be the preferred method on most of these locks due to its ability to get a better bite behind the cylinder.

These locks may be found below the normal entry lock and door knob to prevent someone from kicking in the lock.

FORCING THE TUBULAR DEAD BOLT

Problems Encountered When Using The K-Tool

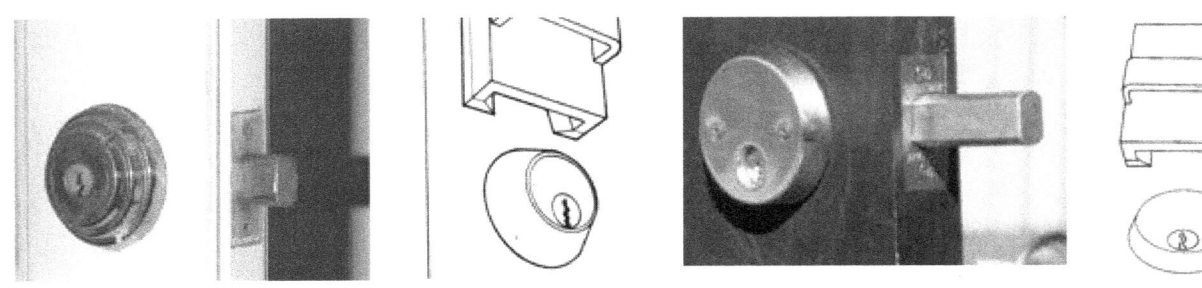

Cylinder Too Deep   **Cylinder Too Wide**

RIM LOCKS

These locks are usually installed as an **add-on lock.** They are installed on the **inside surface of the door** (with the cylinder extended through the door). Only the cylinder is visible from the outside of the door. See Chapter 5 for types of rim locks.

PRINCIPAL OF OPERATION – RIM LOCK

The back of the rim cylinder has a stem, which is inserted into the backside of the lock.
As the key is rotated in the cylinder, **the stem** on the back end of the cylinder rotates the latch or bolt, which locks or unlocks the lock.

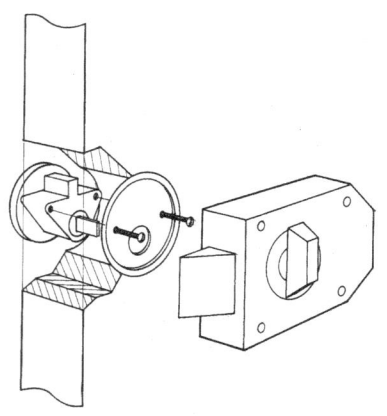

FORCING A RIM LOCK

Using A Lock Puller (Officer's Tool / K-Tool)
- Set the lock puller behind the cylinder getting a secure purchase.
- **Pry up** on the lock puller, pulling the cylinder from the door.
- The back plate is either pulled through the opening or the set screws are ripped from the back plate.
- Insert correct **"Key Tool"** and turn, unlocking the lock.

Note: The cylinder is held in place by two set screws through a back plate. It is the back plate being pulled through the cylinder hole that determines the difficulty.

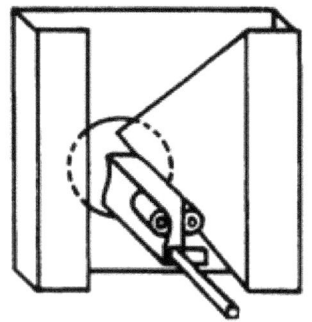
Set the Tool

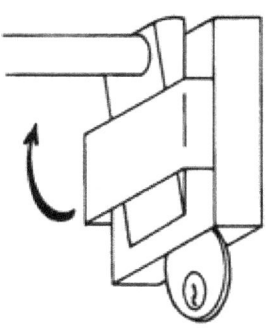
Pry the Cylinder Up

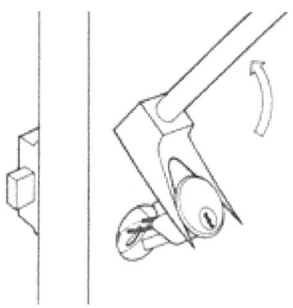
Back Plate Pulled Through

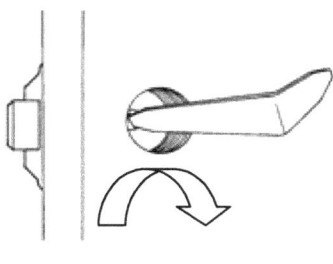
Turn Key Tool

Note: Once you have pulled the lock cylinder be sure to use the proper end of the Key Tool.

FORCING A RIM LOCK

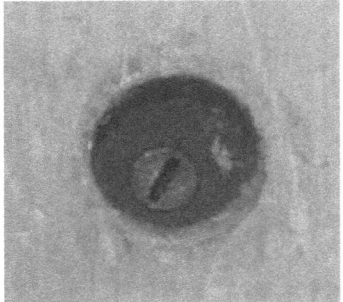
Lock Cylinder Removed

Proper End of Key Tool

On some rim locks, a "shutter" may be installed over the lock mechanism. This will prevent the insertion of a Key Tool. You may have to drive the lock off the door with the tool inserted in the cylinder hole.

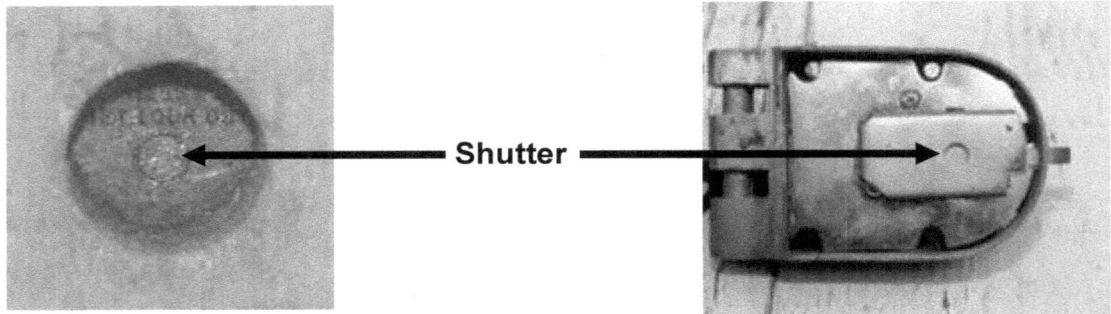

Driving Off the Lock

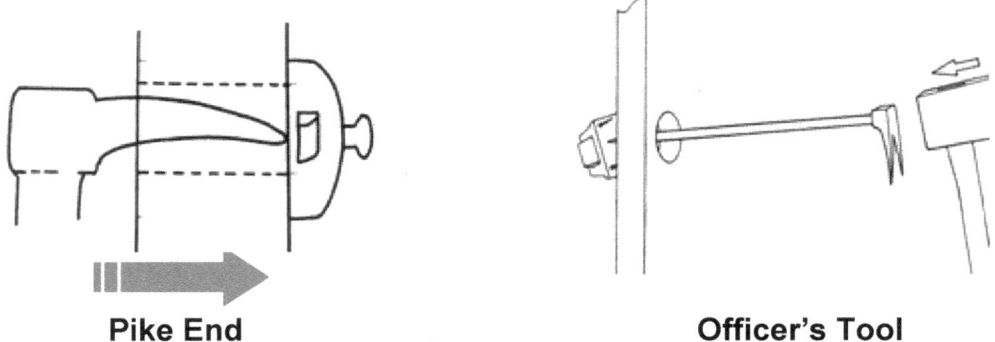

SPECIAL TYPE RIM LOCKS

POLICE LOCK (Vertical Bar Lock) – Only on *Inward* opening doors

This rim lock is very popular in tenements. It utilizes a removable steel bar which it fits into a slide in the lock and extends into a socket in the floor. The bar usually stays in the slide in the "unlocked" position. Its presence is indicated by the amount of resistance met at the lock. Also, the **cylinder may not line up vertically** with the other cylinder locks on the door since it **does not have a bolt "throw"** like other rim locks. The lock, door and bar arrangement is that of a **right angle**. It **provides resistance** to any forward pressure applied by someone trying **to force the door inward.**

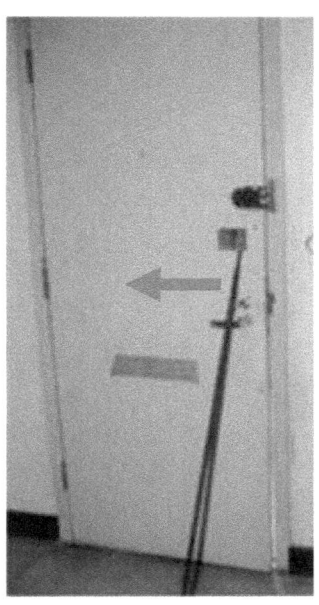

Note: If trying to open from the inside you will have to slide the bar toward the middle of the lock. You may need a tool to knock the bar from the keeper.

FORCING THE POLICE LOCK – Using Officer's Tool / K-Tool

Recognize the lock that you are dealing with. Generally it will not be in line with the other cylinders since it has no bolt or latch (throw) into a keeper. The key in forcing this lock is the **recognition** of the type of lock.

FORCING THE POLICE LOCK – Using Officer's Tool / K-Tool

- Using a Lock Puller (K-Tool, etc) **Pull the cylinder.**
- **Insert and turn** the correct "Key" tool. Note the Key Tool is turned the same as a key.

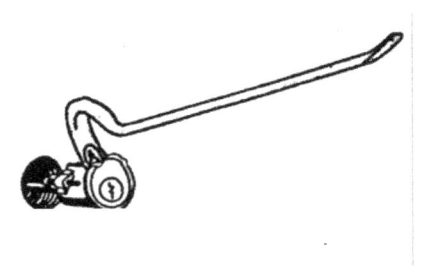

Pull the Cylinder

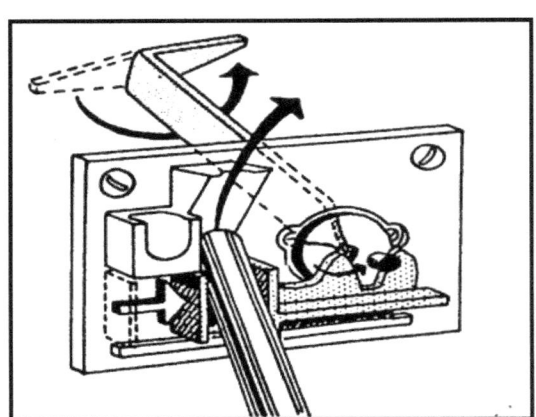

Turn Key Tool

Note: The Key Tool is turned the same as a key. As the Key Tool is turned, the cam slides over moving the bar out of the slot.

Conventional Forcible Entry – Police Lock

If the Thru-the-Lock method does not work, you will have to resort to a conventional forcible entry method. This may not be an easy task.

- **Force any and all other locks** first, e.g. rim and mortise locks.
- Sharply **"Batter"** the Police Lock with the Halligan Tool or axe, attempting to bounce the bar out of the floor slot.

Conventional Forcible Entry – Police Lock

- **Pull the knob** toward you and strike the bottom panel of the door causing the bar to jump out of the floor slot.
- After removing the cylinder, drive the lock off the door with the pike of the Halligan Tool or the Officer's Tool.
- As a last resort, **knock the panel out** of the door, but **make sure there is a charged line in position.**
- **Forcing the Hinge Side** of the door. **Make sure there is a charged line in position,** and try to keep the lock side open as far as possible to make it easier to force the hinge side. Force the **upper** hinge first.
- **BARGE** the door in with your shoulder to bend the bar. This will take considerable force and should be done with at least two men.

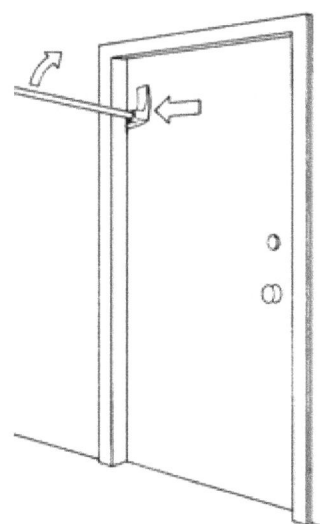

Note: **The last two methods destroy the integrity of the door and a charged line in position is a must.**

FOX LOCK (Double Bar Lock)

This rim lock is easily recognized from the outside of the door by the location of the **lock cylinder in the center of the door.** The cylinder is usually shielded by a rectangular metal plate, which is held in place with four carriage bolts.

Two sets of carriage bolts near the outer edges of the door will indicate the location of the cradles, which guide the locking bars.

The handle in the center of the lock must be pulled toward the operator to engage the clutch inside the lock to move the bars.

FOX LOCK (Double Bar Lock)

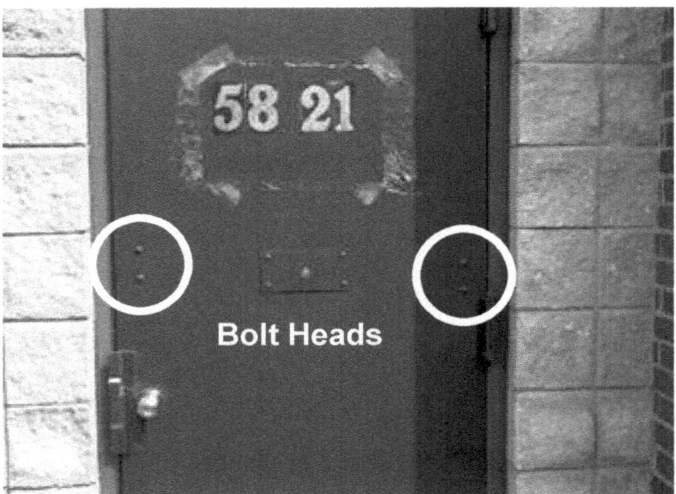

Outside View

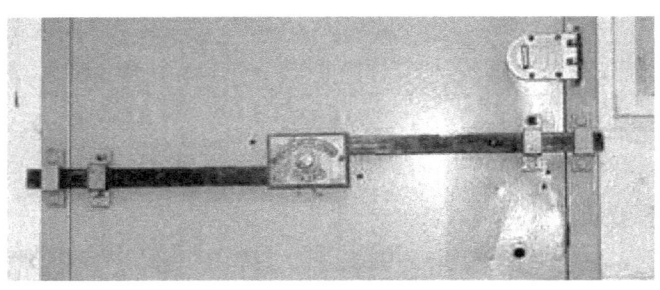

Inside View - Locked　　　　　**Inside View - Unlocked**

Points to consider:
- The rotation of the key turns a gear between the sliding steel bars. This slides the bars in and out of the doorframe. These bars can penetrate **up to two and one-half inches** inside the doorframe or into brackets mounted on the frame itself.
- The position of the bolts, which guide the bars into the doorframe, will show you which way to turn the key. **Usually turning toward the lower set of bolts will unlock the device.**
- The **most effective** tool to pull the cylinder is a K-Tool (lock puller). Also the **ADZ** of the Halligan Tool can be "notched" to catch the inside rim of the cylinder to pull the cylinder, see Chapter 16.

FOX LOCK (Double Bar Lock)

- The cylinder of this lock is generally **recessed or flush mounted**, which initially makes the "K-Tool" ineffective for pulling the cylinder. The door may have to be battered and dented slightly to make a purchase. At times, by using the **PIKE** of the Halligan Tool, you can pry the cylinder out.
- After pulling the cylinder, and inserting the Key Tool, exert a little **inward pressure on the Key Tool.** This will push the gear into the teeth of the locking device, thus engaging the drive gear. **Maintain this pressure** and turn the "Key Tool" to unlock the device.
- To open this device from the inside **the thumb knob must be pulled toward the operator** to open the lock.

Note: Always look for the door the occupant uses as the MAIN ENTRANCE. Secondary doors can be very formidable and difficult to open even from the inside.

FORCING THE FOX LOCK

- **Recognize** this lock.
- **Shear** three of the bolts holding the cylinder guard (plate) and **PIVOT** the plate to expose the cylinder.
- **Remove** the cylinder, using the K-Tool, Officer's Tool, modified adz end or pike end of Halligan.
- **PUSH** and **TURN** the Key Tool (5/32-inch square stem) **IN** to engage the clutch, usually toward the lower bolts holding the bar, same as a regular key.

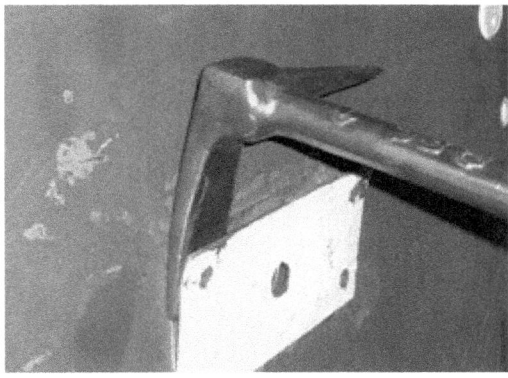
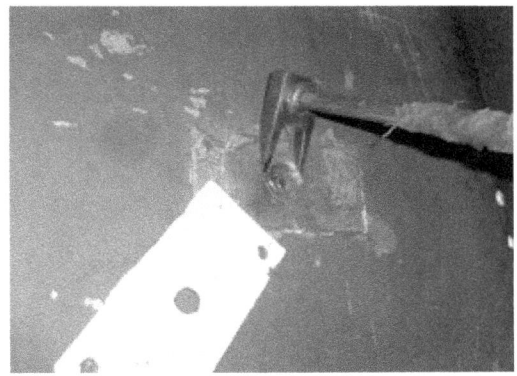

FORCING THE FOX LOCK

 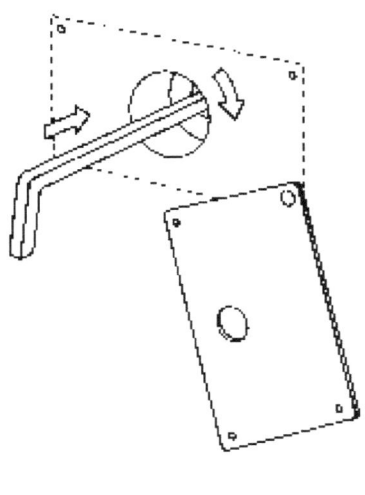

Note: Be sure to use the proper Key Tool (5/32 square).

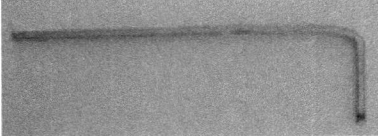

MORTISE LOCKS

Are designed and manufactured to fit into a cavity in the edge of the door (either metal or solid wood). They have a solid, threaded key cylinder which is held in place by two set-screws. There are various types and styles of these locks available today.

A deadbolt and latch is a mortise type lock that contains both a latch and a bolt in one unit.

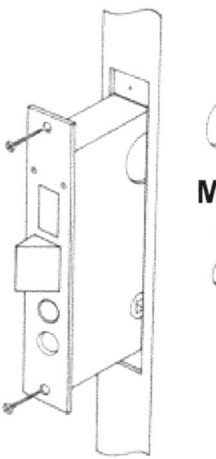 

Mortise Lock

Set-Screws

PRINCIPLE OF OPERATION – MORTISE LOCKS

As the key is rotated in the cylinder, it turns a cam on the back of the cylinder. This cam makes contact with a lever inside the lock box removing it from the strike. Although the key will cause the cam to make a complete revolution, the actual work of opening the bolt is usually accomplished between 5 and 7 o'clock or 7 and 5 o'clock of that revolution depending on which side (right or left) of the door the lock is mounted.

FORCING THE MORTISE LOCK:

- Set the K-Tool firmly on the cylinder and remove the cylinder by **pulling up.**
- Insert the correct Key Tool.
- Rotate the Key Tool. If the mechanism is found at 5 o'clock, rotate toward 7 o'clock, if found at 7 o'clock, rotate toward 5 o'clock.
- If mounted with a doorknob, it may have a latch that may be connected to a second assembly. This may **necessitate a second revolution** of the cam to remove the cam from the keeper.

Note: This second revolution may start a little higher in the opening, e.g. 9 o'clock or 3 o'clock.

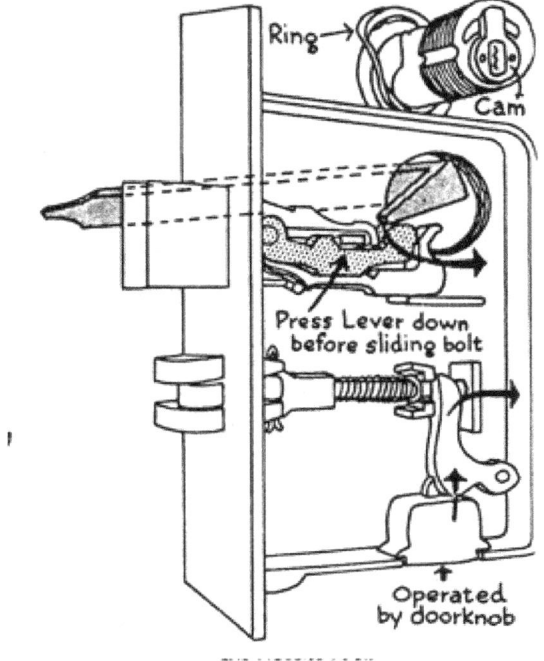

Note: Once you have pulled the lock cylinder be sure to use the proper end of the Key Tool.

Lock Cylinder Removed

Proper Key Tool

PIVOTING DEADBOLT

This popular lock is usually found on aluminum and glass panel doors with narrow stiles. It is also found on solid glass (tempered glass) doors with the frame on the top and bottom edges only. Generally these are commercial occupancies.

PRINCIPLE OF OPERATION – PIVOTING DEADBOLT

These particular locks usually have a laminated bolt, which may extend up to 1-3/4 inches. The tripping mechanism is slightly different than other mortise locks, which requires the correct Key Tool to be used to depress the locking pin, which rotates the dead bolt. The pivoting bolt allows forward throw to be the entire depth of the frame channel.

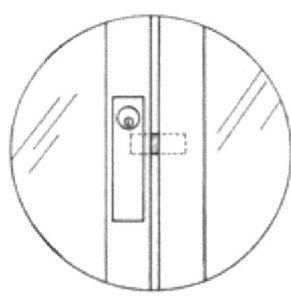

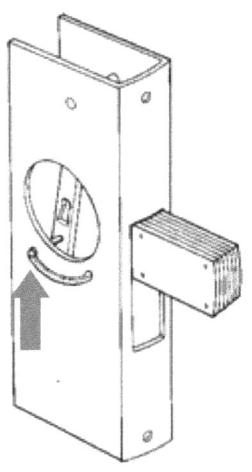  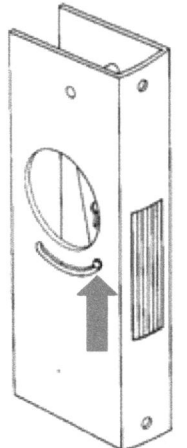

Pin Away   **Bolt Pivots**   **Pin Forward**
Door Locked   **Into Frame**   **Door Unlocked**

The above is a narrow stile, pivoting deadbolt showing the 1¾ inch laminated bolt. The locking pin is **AWAY** from the leading edge of the door. The door is **locked** when the pin is in this direction. As it is depressed the bolt **"pivots"** into the frame. When the locking pin is **FORWARD**, the bolt is inside the frame and the door is **unlocked.**

FORCING THE PIVOTING DEADBOLT - Using The K-Tool

This particular device is virtually impossible to force conventionally (axe and Halligan) without breaking the glass insert and or destroying the door and or the frame because of the long **throw of the DEADBOLT (up to 1 3/4 inches).**

Pulling this cylinder is usually no problem for the **"K-Tool"** (it was designed for this lock). These cylinders rarely break apart.

FORCING THE PIVOTING DEADBOLT - Using The K-Tool

- Place the K-Tool over the cylinder and set by driving down over the face of the cylinder to lock onto the cylinder.
- Pry **UP** with the **ADZ** end of the Halligan, removing the cylinder.
- Using the bent end of the Key Tool, **DEPRESS** the pin and **SLIDE** the pin forward, pivoting the deadbolt down into the housing.
- As the locking pin slides forward, the bolt is retracted, unlocking the door.

Note: The pin will be located at either the 5 o'clock or the 7 o'clock position. Move the pin from 5 o'clock to 7 o'clock, or from 7 o'clock to 5 o'clock to retract the bolt, unlocking the door.

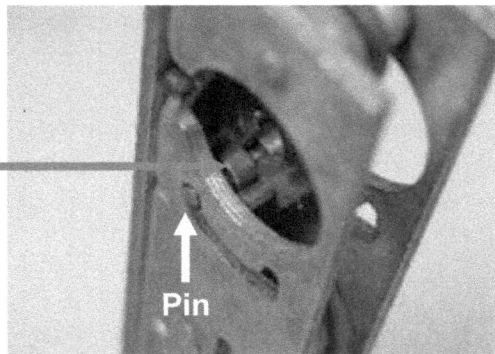

Place End of Key Tool Here to Depress Pin

FORCING THE PIVOTING DEADBOLT – Using The Visegrips

The cylinder may be able to be turned out of the lock by using a pair of visegrips. Since all cylinders are held in place with set-screws, a quarter turn clockwise may bend the set screw just enough to allow you to **turn the cylinder counter-clockwise** and remove it. After entry is accomplished, the cylinder may be screwed back into the lock box.

This method may work on most mortise locks.
If the cylinder guard is beveled or rotates freely, pulling the cylinder is a difficult, if not impossible, task.

FORCING THE PIVOTING DEADBOLT – Using the Visegrips

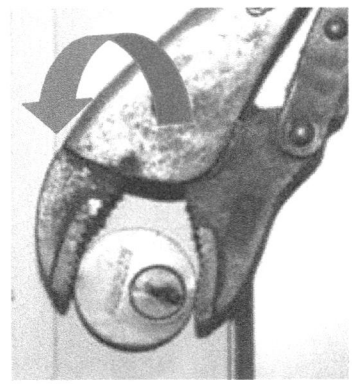

Visegrips On Cylinder   **Unscrew Cylinder**

Note: At no time do we recommend breaking the glass in the door. The reason is safety. If it is "Plate" glass, the broken pieces may be quite large and very heavy. If they are in front of the doorway, they can become a tripping or slipping hazard. Glass and water make for a very unsafe combination when on the ground.

If a glass piece hangs up in the frame it may become dislodged and strike a member causing a severe cut or laceration. This is quite common since the smoke coming from the occupancy may cover the upper portion of the doorframe obscuring any fragments left in the door.

ALTERNATE MEANS OF FORCING – Using The Saw

If the occupancy has **center opening double doors,** take the **forcible entry saw** with the metal cutting blade and cut the bolt between the doors. There is enough room between the doors because of the door swing and the space is usually covered with only weather stripping. This may work with a single door if there is clearance for the saw to get in.

ALTERNATE MEANS OF FORCING – Using the Saw

Note: If the bolt has a ceramic insert it will be more difficult to cut through.

ALTERNATE MEANS OF FORCING – Using the Halligan

Place the **ADZ** end of the Halligan between the door and the jamb, with the bar of the Halligan in line with the cylinder, and parallel to the ground. Strike the Halligan with an axe or maul **DOWNWARD** on the **ADZ**. This may snap the pin holding the bolt and pivot the bolt out of the keeper.

This may work with single or double doors as long as there is room to place the Halligan.

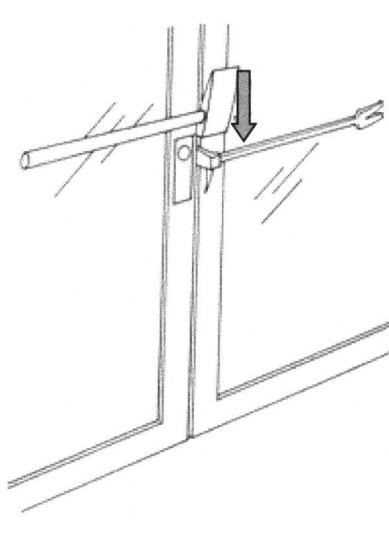
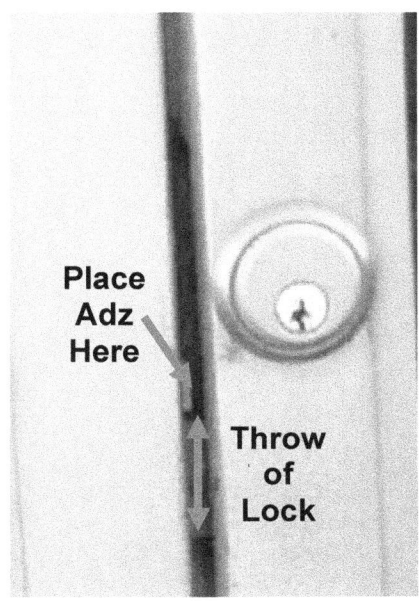

PADLOCKS

PADLOCKS

Padlocks are detachable locking devices having a sliding and pivoting shackle that pass through fixed or removable hardware and then made secure.

This chapter provides information and recommended procedures and tools used for forcible entry of padlocks. Like any fire or emergency, operational procedures and conditions on arrival will dictate the course of action. Is it a *tactical response* - Fire and/or life threatening emergency or is it a *routine response* - Non-life threatening emergency.

Padlocks are used in all types of occupancies, e.g. multiple dwellings, commercial, private dwellings, vacant buildings and even subways and railroads.

Padlocks are used on both the exterior and interior of occupancies. They are found in the places you least expect and you may have to force one with only the tools you carry. Therefore, members should be able to identify the various types of padlocks and their attachment hardware and means of installation.

For the purpose of this manual, the names of the locks used by the author are "street" names and not the manufacturer's product name.

CATEGORIES OF PADLOCKS
For the purpose of size-up and understanding of padlocks, they are placed in three (3) categories:

- Light duty.
- Heavy duty.
- Special purpose.

PADLOCK SIZE UP:
- Type of padlock.
- Hardware and installation (attachment device).
- How many padlocks and their location.
- Accessibility.

PARTS OF A PADLOCK
- Shackle or bow.
- Body, solid or laminated.
- Keyway.

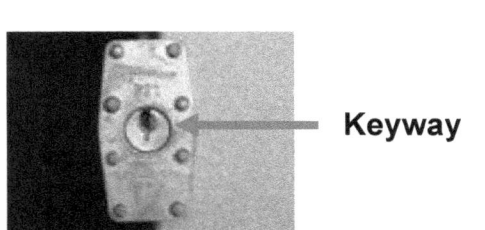

LIGHT DUTY PADLOCK

- Shackle or bow is usually **1/4 inch or less.**
- Shackle or bow **usually not case-hardened.**
- Body of lock, solid or laminated.
- Keyway (type may vary).

HEAVY DUTY PADLOCK

- Shackle or Bow, **1/4 inch and larger.**
- Body of lock, solid or laminated.
- **Case-hardened steel.**
- Toe and heel locking.
- Guarded keyway.

SPECIAL PADLOCKS

Hockey Puck / American 2000 Series

Round padlock, American 2000 is the most common. May also be called a "doughnut" lock.

- No exposed shackle.
- Locking device fits over the staple.
- Removable pin.
- May be case-hardened.

SPECIAL PADLOCKS

Fasco Lock
This is a heavy-duty, surface mounted slide bolt that is locked with an Amercian 2000 lock. The Fasco has a built-in guard to protect the lock. It is bolted to the door with ½ inch carriage bolts, which makes shearing the bolts impractical. This lock is usually found on doors in maintenance rooms in housing developments.

Horseshoe Padlock
- "U" shaped body made of **brass or case-hardened steel.**
- Exposed **"straight" traversing shackle.**
- Guarded keyway.

Wrapped Lock
Constructed on an individual basis, it is not a commercially sold padlock and will vary in strength.
- Heavy **gauge steel welded** to the lock.
- **Limited access** to the keyway.

GATE LOCKS

These are devices made specifically for securing roll-down security gates. There are a few varieties of gate locks that are becoming very popular in urban areas. Here are a few of the most common the authors have encountered.

Bolt Lock (Medeco)
This case-hardened, tubular steel device goes through the gate and rail securing the occupancy. When secured properly it is very effective.

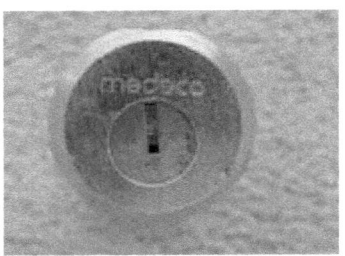 

Mushroom Lock
This device is secured into the bottom rail of the security gate.

Tank Lock
An extreme method of protecting the padlock. This is manufactured on the site. It is steel welded to the frame protecting the padlock.

More On These Locks In Chapter 14.

ASSOCIATED HARDWARE USED WITH PADLOCKS:

Hasps
Manufactured in many different sizes and strengths. They may be in-stalled with screws or bolts, which may be guarded by the hinge while in the locked position.

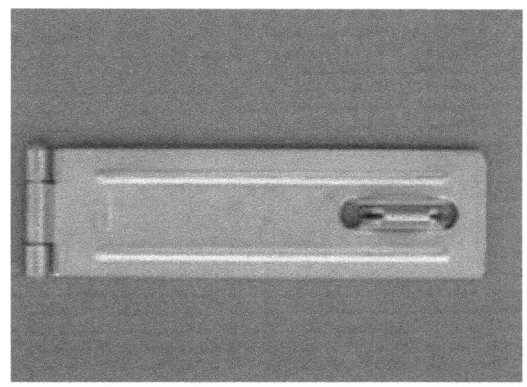

Slide Bolts
A device that travels in a track, which locks into a recessed hole or hardware. Padlocks pass through rear of bolt and are made secure. These slide bolts may be made of case-hardened steel. They are installed with screws or carriage bolts which may be exposed or guarded.

Note: **An alternate means of forcing a slide bolt is to place the FORK or PIKE end of the Halligan Tool inside the shackle and twist the entire lock to break the hardware (slide bolt) holding the lock.**

FORCING PADLOCKS – Using The Forcible Entry Saw

Use the aluminum oxide blade. This should be the **primary tool** to remove padlocks, hardware and attachment devices. It offers speed and is relatively safer than striking tools.

Padlock with *Exposed* Shackle:
- Rotate the padlock to get a cutting position.
- Cut through **BOTH SHACKLES AT THE SAME TIME.**

Padlock with *Shielded* Shackle

This could be the American 747 series or a wrapped lock.

- Rotate the padlock; confirm that the shield covers both front and rear of the lock.
- Cut through both shields at same time.
- Apply two vertical cuts through the shackle if accessible.

Hockey Puck Lock (American 2000 Series)

Cut through the body of the padlock **3/4 up from the keyway.** If lock remains engaged after being cut through, strike the side of the padlock with a sharp blow. This will usually remove the lock. Some of the newer 2000 series have a shielded keyway which must be cut to open the lock.

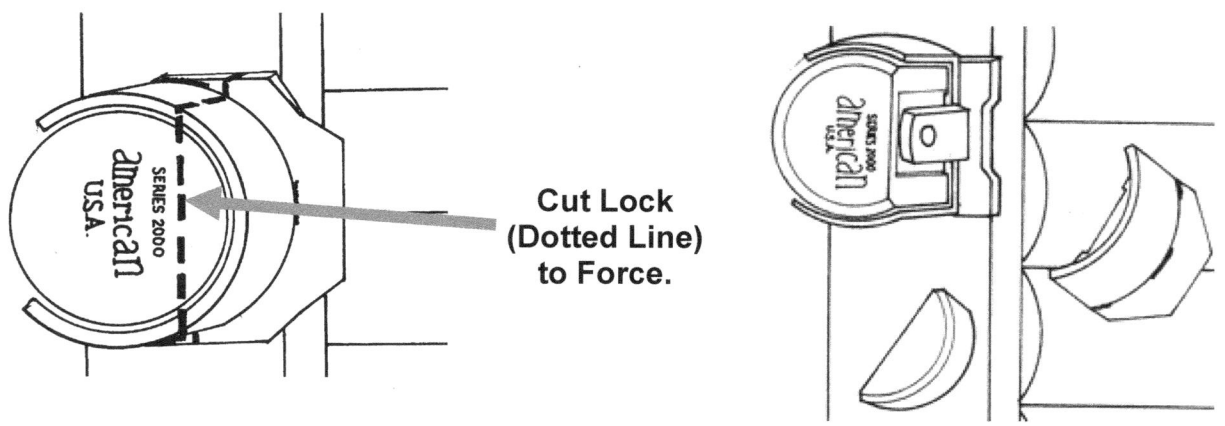

Cut Lock (Dotted Line) to Force.

FORCING PADLOCKS – Using the Power Saw

Note: THIS IS A CHANGE FROM TRADITIONAL CUTS OF 2/3's UP THE KEYWAY. YOU MUST CUT ¾'s UP FROM THE KEYWAY TO CLEAR THE INTERNAL HASP.

Horseshoe Padlock
- Cut through the body of the lock and the shackle.
- At times it may be necessary to make two cuts, one at each end of the shackle.

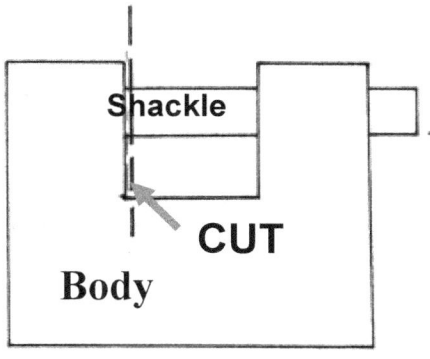 

Fasco Lock

METHOD I - Cutting the Slide Bolt
- Cut through the slide bolt.
- Hold saw at an angle to cut a wedge out of the guard.

METHOD II - Cutting the Lock
- The lock can be forced by cutting the body of the lock ¾ up from the key cylinder.
- Cut the lock until the key cylinder slides out of the body of the lock.
- Remove the lock, slide the bolt.
- If the key cylinder does not slide out, cut until the lock is in two pieces. Push the pin out and remove.

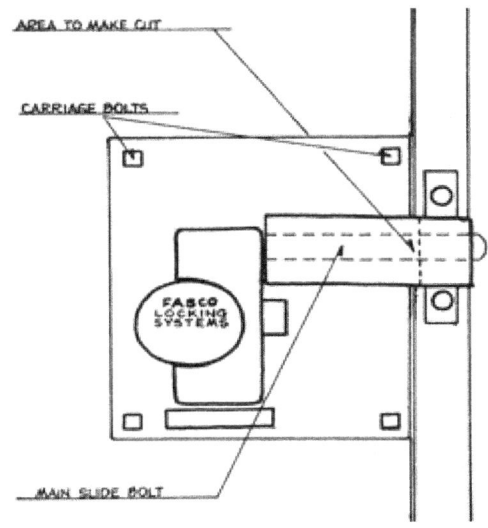

FORCING PADLOCKS – Using the Power Saw

Fasco Lock – METHOD I - Cutting the Slide Bolt

Cut at an Angle

Cut a Wedge

Cut through Bolt

Fasco Lock – METHOD II - Cutting the Lock

Cutting the Lock

Key Cylinder Popped Out

Note: The location and position of the padlock may not allow the power saw to be used.

FORCING PADLOCKS – Using the Duckbill Lock Breaker and Pike of the Halligan Tool

DUCKBILL LOCK BREAKER

Place the Duckbill point between the shackle and the body of the padlock. Keep the tool in line with the padlock. The position and location of the padlock will determine the difficulty of the operation, e.g. padlock is too high or too low. Strike Duckbill with an axe or maul.

Members Holding the Duckbill Tool Should:

- Keep both hands on the tool with a firm grip.
- Keep the Duckbill tool as vertical as possible.
- Keep their eyes on the tool.

Members With the Axe or Maul Should:

- Position themselves where they can deliver the maximum force to lock.
- Strike the Duckbill perpendicular to the head.
- Start with short chopping blows until head is set into the shackle.
- More powerful blows are delivered until the lock body and the shackle are separated.

SAFETY NOTE: **During these operations, members should be aware that the body of the padlock could become airborne and cause possible injury.**

PIKE OF THE HALLIGAN TOOL

- The **PIKE** of the Halligan Tool may be more effective on padlocks with short shackles.
- Place the pike into the shackle opening, keeping the Halligan Tool as horizontal as possible.
- Maintain pressure on the lock body.
- Deliver sharp blows with a maul or axe.

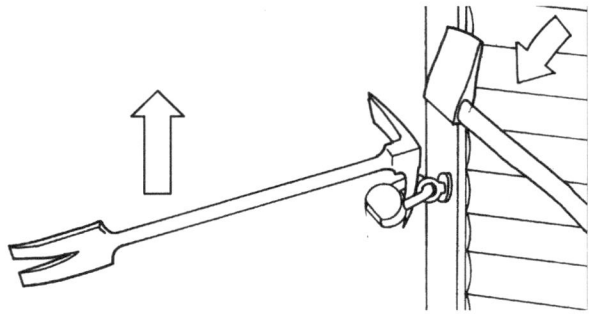 

Technique Tip: **When striking either tool with the axe, the eight pound axe is preferred.**

FORCING PADLOCKS – Using the Bolt Cutters

BOLT CUTTERS

Bolt cutters are excellent for cutting light duty pad locks, light duty chains, cable and hardware. As a last resort they can also be used to cut heavy-duty padlocks, but when used this way, they may damage the jaws of the bolt cutter.

If they must be used for a heavy-duty padlock:

- Open the bolt cutter to the maximum.
- Position the bolt cutter so one handle is securely against a substantial object (wall, ground, etc.).
- Push with both hands on handle to cut the hardware.

Note: Most heavy-duty padlocks have toe and heel locking. Both sides of the shackle may have to be cut, or twisted to remove the lock.

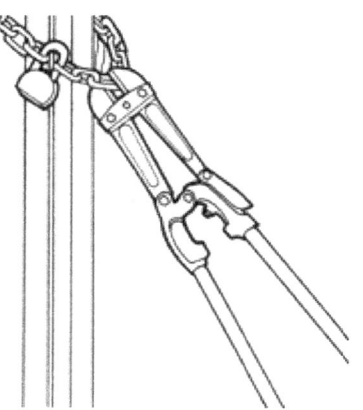 

Cutting the Chain   **Cutting the Lock**

FORCING PADLOCKS – Using the Pipe Wrench

- Secure the jaw over the body of the lock.
- Apply force downward.

Note: This will only work on the American series 2000 lock. This method of forcing this lock WILL NOT work if there is any type of shielding present.

FORCING PADLOCKS – Thru-the-Lock

If you are able to remove the keyway (cylinder), you might expose the lock mechanism and possibly trip the lock using a modified Key Tool or screwdriver. Two methods that have been used with some success are the Bam-Bam tool, which will remove the keyway, and prying off of the guard, (protecting the keyway), allowing the cylinder to possibly drop out.

BAM-BAM TOOL

This is a tool that requires technique, patience and **hardened sheet metal self-tapping screws.**

- Screw the hardened sheet metal screw into the keyway of the padlock **(do not over tighten)**.
- Give the Bam-Bam tool a sharp blow rearward.
- Re-tighten the screw.
- Strike sharp blows until the keyway is removed.
- Insert Key Tool or screw driver and turn to unlock.
- May work with limited success in a **"guarded keyway."**

Note: This device will not work on Laminated locks.

FORCING PADLOCKS – Thru-the-Lock

Hockey Puck / American 2000 Series Lock

- Remove keyway (Bam-Bam tool) or pry off shield exposing keyway.
- Once the keyway is removed, insert rubber tip of Key Tool (modified) into the hole where keyway was.
- Secure onto pin.
- Take a quarter turn and remove the pin.

Shielded Heavy Duty Padlocks

On some models (usually foreign) the guard or shield may be pried out with a screwdriver and the cylinder will fall out. Once the cylinder is out this will expose the mechanism of the lock and it may then be able to be tripped using one of the earlier described methods.

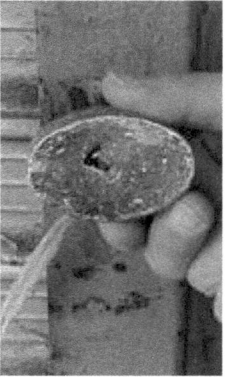

ROLL-DOWN SECURITY GATES

ROLL-DOWN SECURITY GATES

Roll-down security gates are becoming quite common throughout many cities. These gates protect storefronts, factories, warehouse and residential occupancies. They are also used to secure occupancies **inside buildings**, vacant buildings and roof bulkhead doors in vacant buildings.

Adjacent to the opening (window or door) two channel rails are secured to the exterior wall. These are known as the "**guide rails**." Above the guide rails is a drum which houses the **curtain** (interlocking slats of metal or fiber glass). The **slats** ride up and down in the guide rail covering the opening. The curtain may be raised manually, mechanically (with a chain assist) or through electricity. All roll-down gates are constructed the same, except for the opening mechanism.

FIRE GROUND PROBLEMS

Designed for security, they have added to our fire ground problems by:
- Delayed discovery.
- Intense fire upon arrival.
- Extension of fire throughout.
- Very high heat and heavy smoke condition.
- Potential for back draft.
- Ventilation delayed and limited.
- Potential for wall collapse.
- Difficulty in locating the seat of the fire.
- Time consumed in extended forcible entry.
- Need for power saws to gain entry.
- Difficulty in determining the exact entrance door, when numerous gates are present.
- May block entrance to sidewalk cellar door, upper floors and FD Siamese connection.
- Gates may be secured from the inside, and occupants use another exit to leave building or worse yet, lock themselves inside.

TYPES OF GATES:

- Sliding Scissor Gate.

- Manual Roll-Down Gate.

- Mechanical Roll-Down Gate.
    - Chain Operated
    - Gear Operated

- Electric Roll-Down Gate.

SLIDING SCISSOR GATE

This is the oldest type of security gate. These are among the first barriers that owners put in place to discourage vandalism and break-ins. Unlike the more common gates we encounter today, **these gates slides in a track to open.**

Construction Features

- The bottom track usually picks up and secures the gate in the open position; some pivot ninety degrees to achieve the maximum opening.
- These gates may be secured with numerous padlocks.
- These locks will be located in the center of the opening of the gate cover or off to one side, attached to the frame.

Forcible Entry Operations

- Locate and remove all padlocks and / or other locking devices.
- Slide the gate manually.
- Lift the bottom track and secure in open position. If possible, rotate gate ninety degrees to achieve maximum opening.

MANUAL ROLL-DOWN GATE

Usually found on the front of smaller occupancies. These gates can cover an entire storefront or just a doorway.

Construction Features

- Gates ride up and down a channel rail on each side of the gate.
- The slats may be wider on the older gates.
- The gate is attached to a winding drum.
- At the top of the gate (on larger manual gates), the drum may have a spring counter-balance to assist in the opening.
- The winding drum is concealed behind sheet metal housing or inside the building wall.
- These gates are secured with metal pins that pass through the channel rail and the gate. These pins are secured to the channel rail with a padlock that attaches to a metal clip or staple welded to the channel rail.
- Each gate may be secured with numerous padlocks.
- The manual gate is easily recognized by the absence of a raising mechanism housing on the side of the winding drum (top of the gate).
- Lifting handles are usually attached to the bottom rail of the gate.
- Slide bolts may be attached to bottom rail and may be secured into the channel rail with a padlock.
- The curtain may be constructed of:
  - Inter-locked, solid sheet metal slats.
  - Open grill metal bars, connected with metal tabs.
  - Fiber glass.

MANUAL ROLL-DOWN GATE

Forcible Entry Operations

- Locate and remove all padlocks and/or other locking devices.
- Pull (slide) all metal pins and slide bolts out.
  - Most of the padlock points will have a removable pin.
  - Bottom rail usually has a slide bolt to disengage.
- Raise gate with lift handle or bottom bar.

MECHANICAL ROLL-DOWN GATE (CHAIN HOIST)

All of the same features as the manual gate. These types of gates are generally found on wider openings.

Construction Features
- Gates ride up and down a channel rail on each side of the gate.
- The slats will be narrower, span a wider opening.
- On gates mounted on the exterior walls of buildings, the chain hangs from a narrow metal housing attached to the side of the winding drum housing. The chain is secured behind a hinged piece of angle iron. The chain is attached to a hold-down device such as a bolt to prevent pulling the chain out from the top of the angle iron. The angle iron is secured to the channel rail with one or more padlocks.
- On gates mounted with the winding drum concealed in the building wall, the chain will not be visible. The chain will be secured in a small access panel on the building wall adjacent to the channel rail. A key operated latch type lock will secure the access panel.
- The hoisting chain is secured behind a piece of angle iron and usually secured with padlocks.
- Each gate may be secured with numerous padlocks and slide bolts similar to the manual gates.
- The gate is usually larger, hence heavier.

MECHANICAL ROLL-DOWN GATE (CHAIN HOIST)

Forcible Entry Operations

- Locate and remove all padlocks and/or other locking devices.
- Pull (slide) all metal pins and slide bolts out.
- Free the chain hoist from its hold-down device and raise the gate with the chain.
    - The angle iron covering the chain hoist is usually hinged and has to be pivoted out and away from the rail to access the chain hoist.
    - If the angle iron is not hinged you may have to it pry open to access the chain.
- If the gate cannot be raised with the chain hoist assemblies, cut the chain near the top and raise it manually.

Notice the different housings for the pull chain.

Note: This may take several firefighters since it will be much heavier.

MECHANICAL ROLL-DOWN GATE (GEAR OPERATED)

All of the same features as the manual gate, this is another version of a mechanical gate. The difference in this gate is in how it is raised. The size of the opening is not necessarily an indication if this version of mechanical gate is present.

Construction Features
- Same construction features as the other gates.
- These gates are raised by turning a gear assembly with a crank handle.
- The gear assembly will be located at the top of the hoisting drum, in the same area as the chain hoist version.
- This version of mechanical gate will be able to be determined by what appears to be an eye bolt visible at the bottom of the housing assembly, off to one side.

Forcible Entry Operations

- If the crank handle is not readily available, or housed in the side rail assembly, similar to the chain, cut the gate.

ELECTRIC ROLLDOWN GATE

Same basic features as the other types of gates with the exception that it is operated electrically. It can be found in any occupancy, but usually is found on occupancies with large openings such as department stores and commercial buildings.
  One gate may be used to cover multiple levels of an occupancy.

Construction Features

- Similar to mechanical roll-down but are usually recognized by a large metal motor housing adjacent to the winding drum.
- There may be a key switch located on the building wall on either side of the gate. This switch may also be located in a remote location inside the building. This key switch panel may contain a stop button; others stop with the switch in the center position of the key switch.
- All electrical operated gates are equipped with an auxiliary chain hoist to be used in case of a power failure. This chain will be located in the motor housing and may not be visible from the outside.
- Either a bottom hatch or a front panel, which is secured with sheet metal screws, may access this chain hoist.
- The chain hoist assembly may have a clutch cable or chain that must be pulled first to engage the assembly to open the gate. This electrical gate has now been converted to a mechanical one.

ELECTRIC ROLL-DOWN GATE

- These gates may also be secured with padlocks, pins and slide bolts similar to manual and mechanical gates.

Auxiliary Chain

Forcible Entry Operations

- Locate and remove all padlocks and/or other locking devices.
- If power is ON, operate the electric switch to open the gate.
    - This may be possible in the early stages of the fire.
- Pull the cover off the box to expose the control lever on the back of the switch.
    - Even when the screws are removed, the inner plate must be pried off.

Note: This box may have to be broken to gain access.

These methods of gaining entry may only work if there is **not a large fire or high heat behind the gates.** Once the gate is exposed to high temperatures it may begin to distort and jam as it rides up the winding drum.

ELECTRIC ROLL-DOWN GATE

There is no one, simple method of gaining entry through these obstacles. Each operation has to be treated accordingly. Size-up and the correct tools will dictate the method of entry.

OPEN-GRILL OR DESIGNER GATE (Variation of a Roll-Down Gate)

Same basic design as the other types of roll-down gates. The variation is that sections of the gate, either all or partial, are constructed of small, tubular pieces of metal or metal bars, connected with metal tabs. This type of gate is open so that you may see what is behind it.

This type of gate is generally used where high security is not vital or where the owner wants the public to view the display and also provide some security.

Construction Features

- All the same construction features of the other style gates.
- The curtain may be all or partial of a grid like design.
- Operation of gate will be the same as any other. It can be manually, mechanically or electrically operated.

OPEN-GRILL OR DESIGNER GATE

Construction Features

- Each gate may be secured with numerous padlocks.
- Slide bolts may be present at the bottom.

Forcible Entry Operations

- Locate and remove all padlocks and /or other locking devices.
- Pull (slide) all metal pins and slide bolts out.
- Operate gate based on the type of design, either manual, mechanical or electric.

LOCKING DEVICES FOUND ON ROLL-DOWN GATES

There are many ways to secure these gates. Having some knowledge as to how the locking devices are installed will aid you in removing them.

- Generally there is some kind of opening made into the channel rail and the curtain. Through this opening a "pin" may be inserted which prevents the curtain from moving up.
- There could also be an "eye" buried into the sidewalk to secure the bottom rail.

LOCKING DEVICES FOUND ON ROLL-DOWN GATES

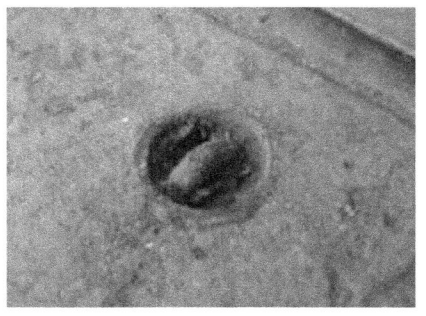  

Steps for Removal:
- Cut or open padlock and remove.
- Remove **PIN.**
- Raise curtain.

GATE LOCK (Bolt Lock)

Another popular device for securing roll-down security gates, fire doors, counter doors and shutters.

Features
- No hasps.
- No shackle.
- No pin.
- Resists cutting, drilling.
- Pick Proof.

GATE LOCK (Bolt Lock)

Forcible Entry

Go for **the weakest point** of this lock, which is the **brass pin which rotates the cam.** One way to force this device is to apply **pressure outward** and try to snap the brass pin.

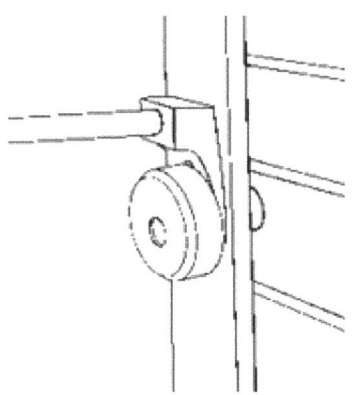

Note: With many of these locks being covered (shielded) with steel, another method would be to try and cut the body of the lock on an angle between the rail and curtain.

EXTERNAL SHIELDS

Most any type of padlock can be found with an external shield. They are used to protect the padlock and to make forcible entry more difficult.

Types of Shielding For Padlocks

- Wrapped shield **welded to padlock.**
- Fixed shield to padlock attachment point.
- Removable shield.
- Welded box.

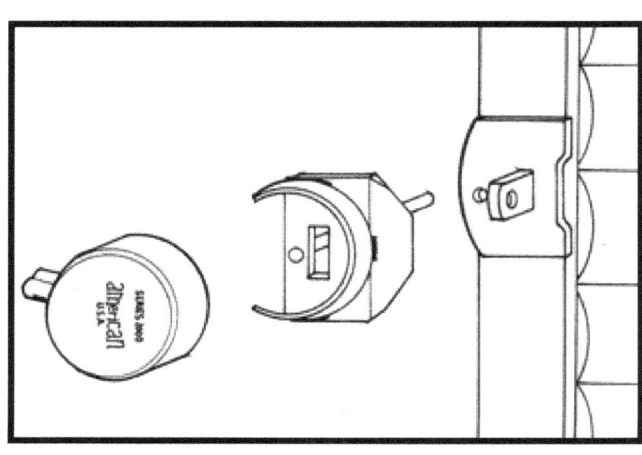

EXTERNAL SHIELDS

Forcible Entry Operations

- Utilizing the Power Saw or Torch, cut through the shield and lock.
- Cut the gate.

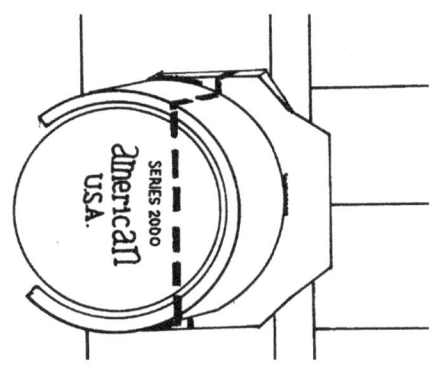

CUTTING THE ROLL-DOWN CURTAIN

There are many ways to cut roll-down security gates to gain access. There are just as many theories to justify these cuts. Each has its own merits but for simplicity, we are showing just a few.

Remember, each fire situation will dictate the appropriate cut.

Note: **Always check to see if the bottom rail is covering a street-level cellar door. Once a cut is made, it will cover over that means of access/egress.**

Triangular Cut

This is the quickest and fastest to get water on the fire. The key to this cut is the overlapping of cuts as **high as you can get and bringing the cuts down to the ground.**

Advantages:

- Only two cuts.
- Ability to put water on the fire immediately.

Disadvantages:

- Large pile of cut gate in front of opening.
- Unable to cut all the way to ground.
- The cut can only be made as high as the saw operator's reach.

  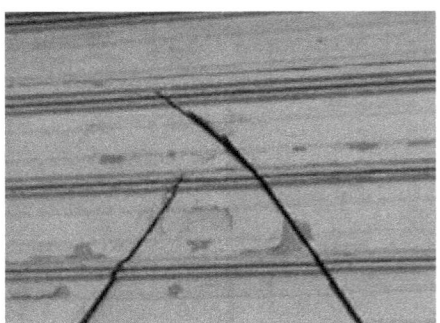

Note: Do not cross cuts at top until the second cut is complete.

CUTTING THE ROLL-DOWN CURTAIN

Triangular Cut

Drive Pike of Halligan into Slat to Remove

Box Cut

This operation requires three vertical cuts. The key to this evolution is ensuring the outside cuts are **at least a foot away from the guide rail.** Again, the cuts have to be made as high as possible and down to the ground. Unlike the triangular cut where gravity brings the cut to the ground, here a couple of slats must be removed manually pulling them from the rail side toward the middle.

Advantages:
- Less of a pile in front of the opening.
- Can be used on very wide openings.
- If done correctly, the opening will be squared.

Disadvantages:
- Requires more time.
- Requires more than one member.
- **Must** remove slat **above** the locking pin.

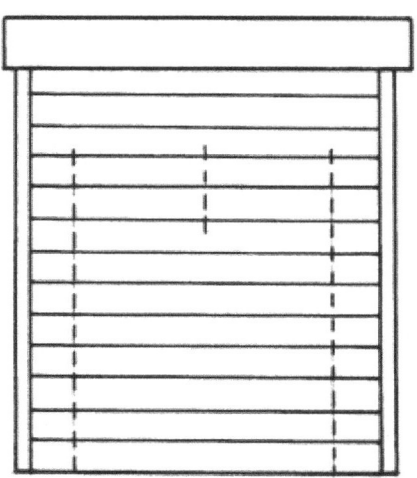

CUTTING THE ROLL-DOWN CURTAIN

Box Cut

Note: Drive the PIKE end of Halligan into slat to remove.
If slats are tight, drive the Halligan with the axe.

OPEN-GRILLE OR DESIGNER GATE

When faced with the necessity to cut the open grill or designer gate a variation of the box cut would be used as shown below.

CUTTING THE ROLL-DOWN CURTAIN

Rail Cut
This opening, making two horizontal cuts in the guide rail above and below the locking device, has been used with limited success. The cut section is pried away from the gate pulling the locking device with it.

Advantages:

- No long cuts into the curtain.
- Faster.

Disadvantages:

- Can only be used where there are only **two** locking devices, one on each rail.

- **May not** pry away with the locking device thereby delaying the operation by then having to do either the Triangle or Box cuts.

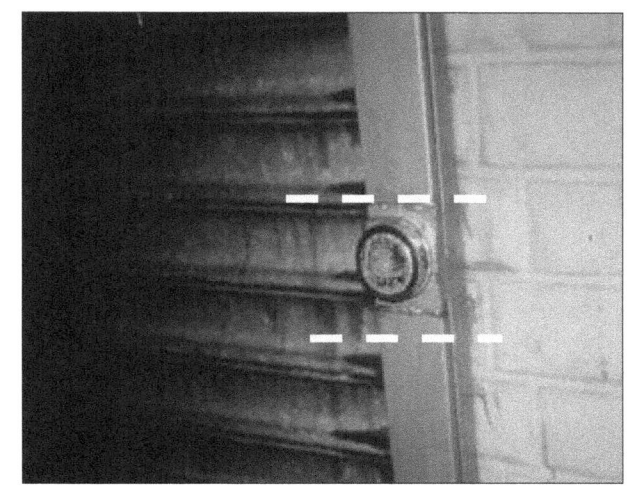

ADDITIONAL TYPES OF LOCKS AND WAYS TO SECURE AND PROTECT

747 Lock   **Covered Lock**   **Earring Lock**   **Long Bow**

ADDITIONAL TYPES OF LOCKS AND WAYS TO SECURE AND PROTECT

Master Lock **Short Bow** **Shielded Lock** **Soda Machine**

Rubber Coated **Different Locks** **Lock Box** **Shield**

 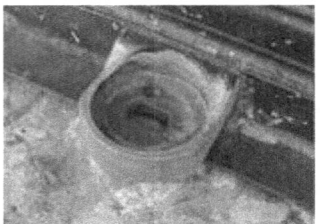  

Bottom Lock Cover **Bottom Lock Holder** **Bottom Rail Lock** **Mushroom Lock**

Horseshoe Lock **Bottom Rail Slide** **Bottom Lock** **Bottom Slide Lock**

ADDITIONAL TYPES OF LOCKS AND WAYS TO SECURE AND PROTECT

 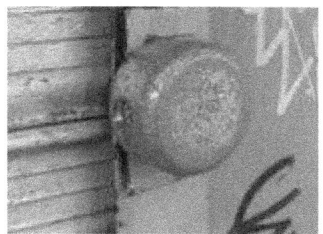  

Non Removal Shield   **No Shield**   **Removal Shield**   **Lock Holders**

Welded Lock Cover   **Lock Cover**   **Static Bar**   **Secured Slate**

MISCELLANIOUS SECURITY PROBLEMS

MISCELLANIOUS SECURITY PROBLEMS

WINDOW BARS

As a rule, bars are fixed and permanent. Gates are designed to open, but may be fixed.

Window Bars
These obstacles come in a variety of sizes, shapes and strengths. They may be mounted to the window frame with screws or bolts, or set into the mortar. Bars and gates are used primarily for security and leave very little room for error in the case of fire.

Attacking and removing these obstacles during a fire situation takes time. If fire is being vented through the window being worked on, it becomes more of a challenge. Anyone trapped behind them has little chance of survival.

Forcing Window Bars

Bars are usually secured to a window at four points. The mounting point may be a lag bolt into the mortar or brick, or the mounting point may be part of the brickwork.

WINDOW BARS

Forcing Window Bars

Bars are attacked at the weakest point. Striking the mortar or brick work where the bar is mounted, may dislodge the anchor point. Dislodge two and bend the bars away, or pry out using the Halligan.

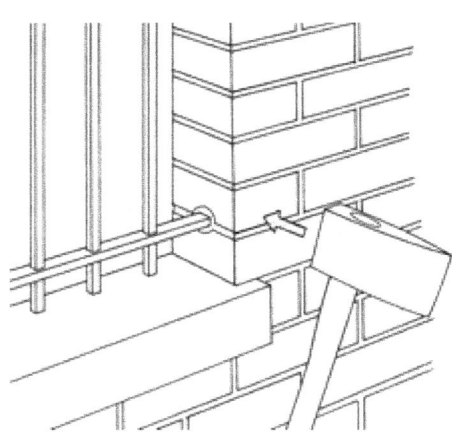
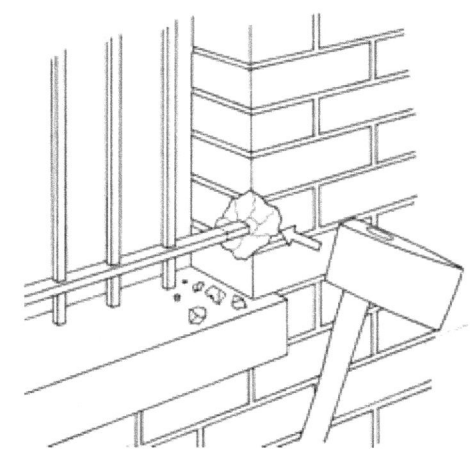

Using a Halligan Tool you may be able to pry the bar from its mount.

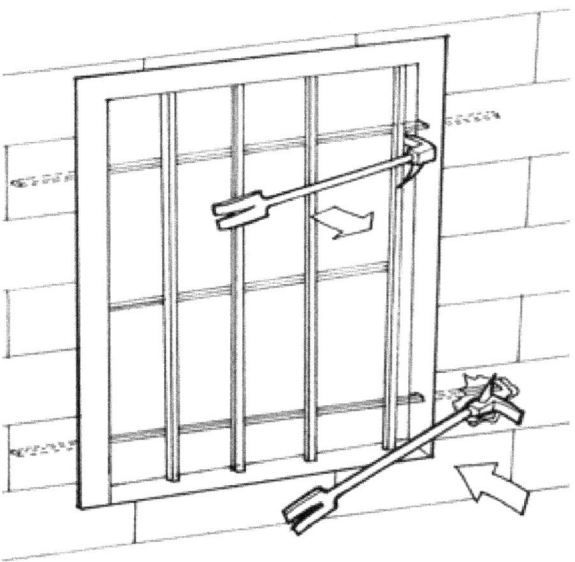

Using a power saw, cut the mounting bracket and remove the entire bar assembly, or cut two sides and bend the bars away.

WINDOW BARS

Forcing Window Bars

Place the Hydra-Ram between the angle iron outer edge of the window bar and the exterior of the building, as close as possible to the lag bolt, and spread.

If the spread is not sufficient, the tool can be repositioned directly at the point of attachment.

After forcing the attachment points on one side, push the gate to the side (while still attached with a hook), allowing an unobstructed opening.
Pushing the gate to the side still attached will cause the gate to break free, dropping it to the ground.
Keep this area clear to prevent any one from getting hit by the falling gate.

Note: Start this operation from the bottom and work up to stay out of the path of the gate if it should fall.

Window Bars: Various Types and Mountings

WINDOW BARS

Window Bars: Various Types and Mountings

WINDOW GATES

Gates come in a variety of types. They vary in size and strength. From the "Scissor" type to the more formidable "Jail-House" type gates.

Gates are usually attacked at the hinge side, since the swing side is usually secured with a padlock. Using the Halligan Tool, the frame of the gate is pulled away from the window frame. The window gate is usually held in place with screws.

In doing this, **the window must be broken.** This will complicate the action because we have "vented" the area we want to access. In gaining access, the entire gate should be removed, as well as the window sash and any window decorations (curtains, verticals, etc.).

Note: When entering via a window with a gate, you have to ensure your way out. Other windows may be similarly fortified.

WINDOW GATES

Approved Type Window Gate

Most people are very concerned about their security and will use any method it takes to guarantee it.

In many municipalities, approved gates may be used. This is to eliminate the need for padlocks or other entrapment devices. These "approved" gates usually have a locking device that is in the form of a lever enclosed behind a small door, thereby eliminating the need for a padlock.

There is no "one way" to remove these obstacles. Like everything else in fighting fires, you have to take what is given you and make the best of it.

Using the correct tool and common sense should get the job done.

Outside View

Release Mechanism

Inside View

Forcing Approved Type Window Gates

The following are suggested methods:

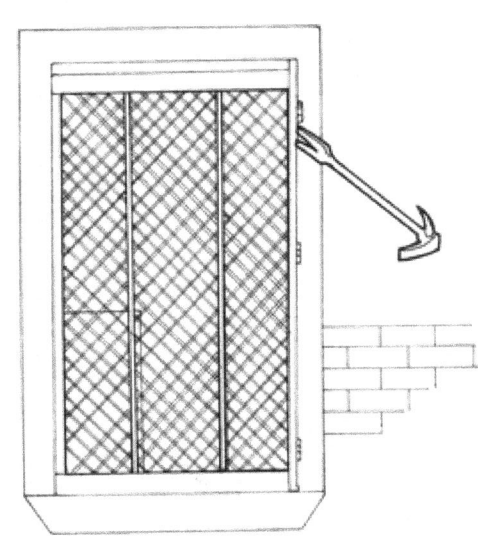

Using the **FORK** end to pry the hinge away from the frame. You will get more leverage this way, but you may be restricted due to a fire escape.

WINDOW GATES

Forcing Approved Type Window Gates

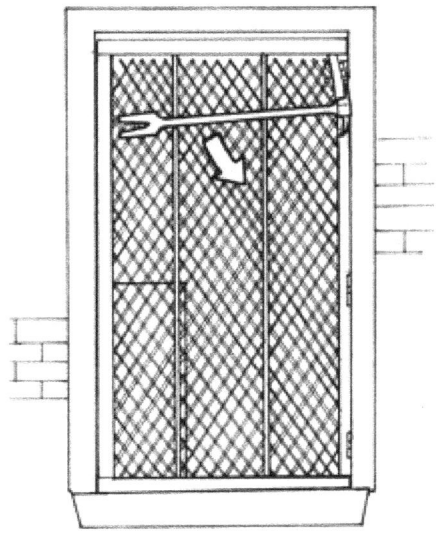 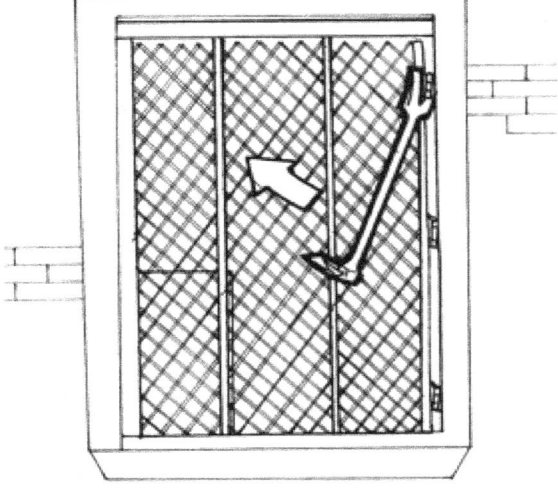

Use the **ADZ** end if there is no room to use the fork end.

Pry off the hinge using the **FORK** end.

Swing Bar Gate (Jail-House Type)

This type of gate is more formidable as it is made of heavy gauge iron bars.

The locking devices are part of the problem as this is not an approved type gate. There is no way to prepare youself for what may be securing these devices.

Swing Bar Gate

  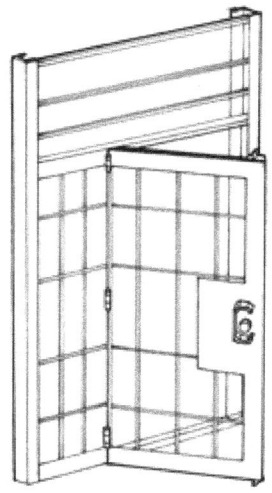

Inside View Closed　　　　　　　　　　　　　　　　**Inside View Open**

WINDOW GATES

Forcing Swing Bar Gate

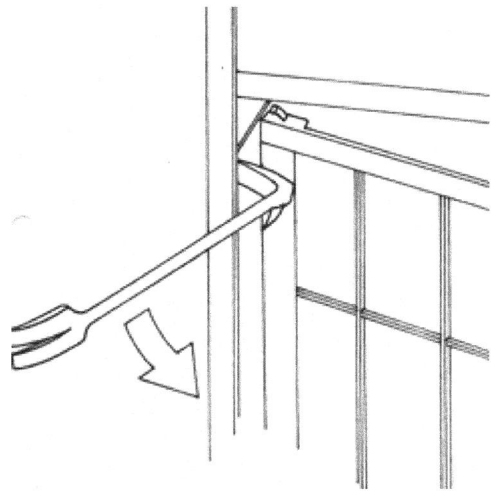

Outside View Forcing Gate

IRON GATES

Usually found at the main entrance. They are mounted like a standard door with hinges on one side and a lock on the other. What makes them formidable is the inability to spread the jamb. They are generally mounted in a metal frame.

If used on a secondary entrance, they are usually more fortified than a simple locking device.

Attacking the lock side in a conventional manner is usually sufficient to gain entry.

CHILD GUARD GATES

Child guard gates come in a variety of weights and sizes. There usually are three to four horizontal bars, which inter-lock and slide to the prescribed opening. The device is secured across the lower sash of the window to prevent children from falling out.

They are usually secured on the outside of the sash to the window frame. They can also be mounted to the inside of the window.

In most cases, screws are used to secure them.

Removing Child Guard Gates

Removing them under **non-fire situations** is usually not a problem. Striking the vertical frame away from the mounting screw will generally be sufficient. In a fire situation, with heat and possible flames, the member may not be able to stand up and swing the tool.

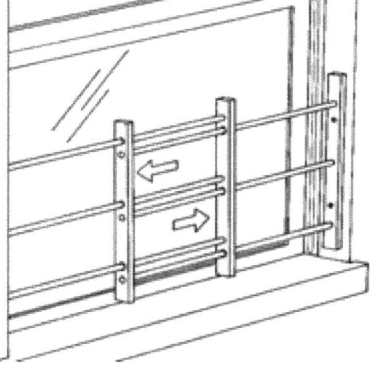

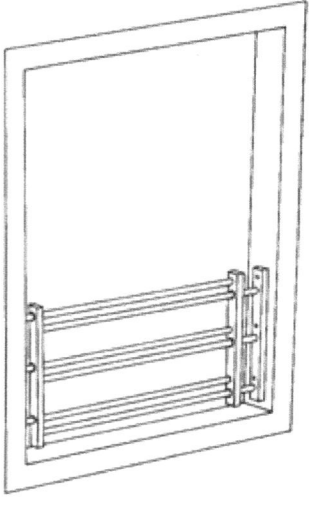

If the gate is larger than the window it was designed for, the vertical frames will be too close to the side of the window frame, preventing placement of the prying tool.

CHILD GUARD GATES

Removing Child Guard Gates

Metal frame windows will be more difficult to force than wood frame windows. Another method of forcing them open would be to strike the horizontal bar where it joins the upright that is screwed into the frame.

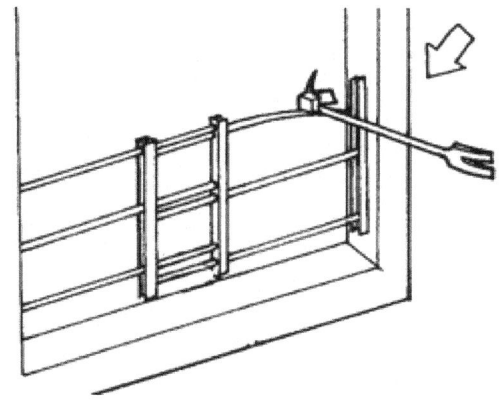

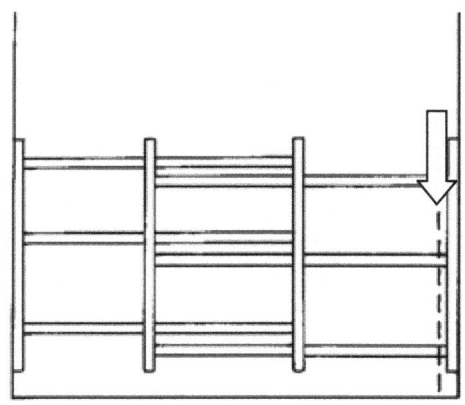

Cutting the horizontal bars is another option, but that calls for a different tool such as a power saw or Sawzall. Bolt cutters may not work.

WINDOW/DOOR BARRIERS

HUD WINDOWS/DOORS

These security devices are relatively new to the fire service. When buildings and in some cases occupancies become vacant, the owner will secure the premises pending further renovation, or re-occupancy.

HUD WINDOWS/DOORS

One or two sheets of plywood over a window opening secured with two or four, 2 x 4's. The 2x4's are wider than the window opening and hold the plywood in place. One or two bolts go through the 2x4's and secure the plywood in place.

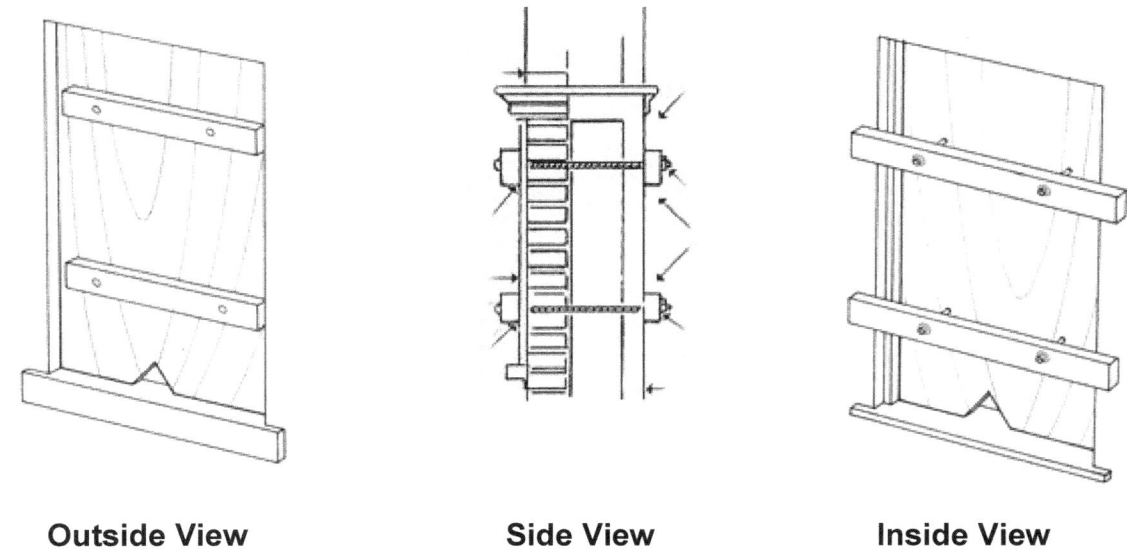

Outside View   **Side View**   **Inside View**

Forcing A HUD Window

To remove this obstacle, using the **PIKE** of the Halligan Tool, strike and split the 2x4 at the bolt. Splitting one side and rotating the other is usually sufficient. This would have to be done to both of the 2x4's.

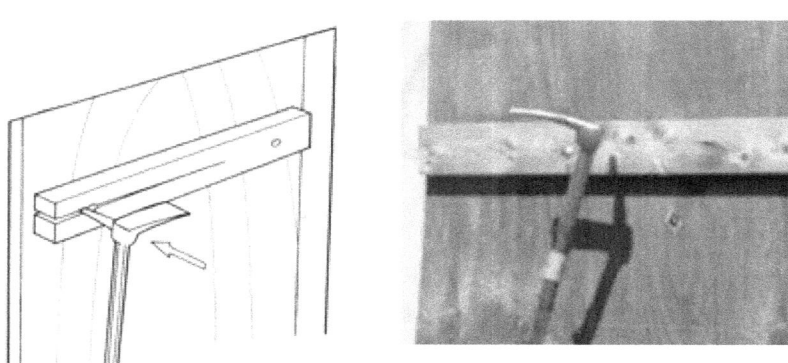

Another method to remove these would be using the forcible entry saw (aluminum oxide blade). In this method, the bolt head that is securing the 2x4 would be cut at a slight angle. Once the bolt heads are removed, you can drive the bolt through using the **PIKE** end of the Halligan.

Note: This method should be used from the tower ladder basket.

HUD WINDOWS/DOORS

Forcing A HUD Window

Note: The danger is in controlling them. If you can not pull the plywood into the occupancy, they may "sail" and anyone in front of the building could be struck.

PLYWOOD COVERING WINDOW/DOOR

When the plywood is secured to the inside of the window frame only, just pry the window trim away and the plywood covering comes with it. If attached on the exterior, this removal is more involved and the same problem exists with the potential for the sheathing "sailing."

Secured Inside   **Secured Outside**

A variation to the plywood covering is nailing wire lath to the sheet of plywood, followed by a "scratch coat" of mortar. This makes an effective seal and offers no point of reference as to where the opening actually is. This method of sealing will usually provide three layers of covering, plywood, wire mesh and mortar, forming an effective seal.

PLYWOOD COVERING WINDOW/DOOR

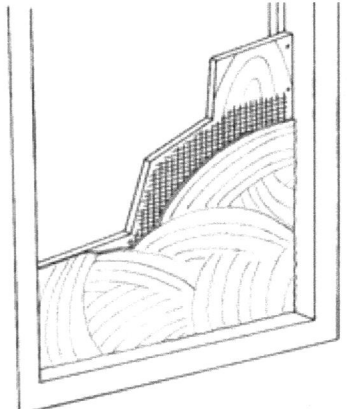  

Three Layers Forming an Effective Seal

Forcing Entry

This type is best removed using the power saw (carbide tip blade). Using one of the following methods will work in most situations:

- Cut an "X" opening in the plywood and peel it back.
- Cut an opening into the plywood, creating another door.
- With a window, cut a triangle for immediate access.

WAREHOUSING

Another method of securing a building or space is called, "**warehousing**." It is very simple and effective, since the contractor may use scrap wood to secure the premises. A piece of plywood is cut to size and secured with screws or nails around the perimeter of the window frame. Then a 2x4 is nailed across the inside of the frame (approximately at the middle of the plywood) with another piece of 2x4 nailed at an angle to the floor, bracing the plywood in position. This effectively creates a right angle.

You will be unable to determine from the outside if the opening is covered in this manner or if it is covered with a piece of plywood secured only to the inside of the opening, until you begin your operation.

The warehousing method of securing an opening is quite difficult to force from the outside.

WAREHOUSING

Outside View

Inside View

Forcing Entry
Once again, the saw is the tool of choice. In this case, a chain saw is very effective.

Note: **When using the chain saw be aware of the depth that you put the bar through when cutting. There may be members inside in the vicinity that could be struck by the rotating chain.**

The door to the apartment may also be secured in a similar manner being covered with a sheet of plywood.
In order to gain entry, the plywood would have to be removed first. Then, whatever is securing the door would have to be forced.

SIDEWALK CELLAR DOORS

This outside entrance into the occupancy may be secured with padlocks. Generally removing the padlock(s) and opening the doors is not a problem.

When the door is **secured from inside or there are no visible locks**, the entire door/hatch should be removed. Strike the mortar around the frame, breaking it away from the frame. Place the Halligan Tools under the frame, lift and then slide the entire door over.

Once the situation is mitigated, slide the door back over the opening and the area is once again secured until repairs to the mortar are completed.

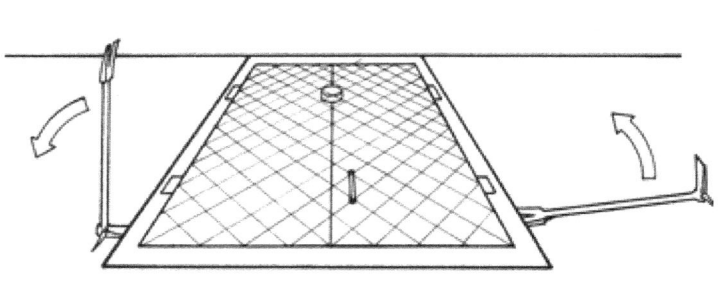

ANGLED CELLAR DOORS (BILKO)

These outside entrances are usually angled from the foundation to the ground. They are generally two hinged steel doors opening from the center. Once open, they give access to a stairway which leads to the cellar or basement. They are secured by a sliding bolt from the underside. They may also have exterior padlocks.

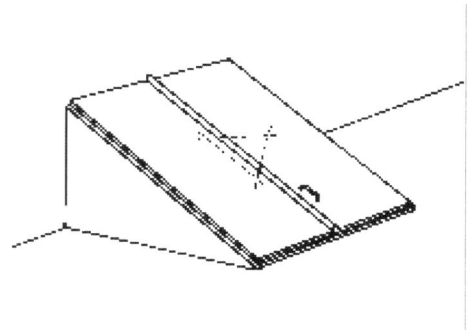

Tools & Methods For Entering Buildings Without Keys

ANGLED CELLAR DOORS (BILKO)

Forcible Entry

- Locate the locking slide bolt, usually near the center of the connecting doors.
- Cut through the door just to the side where the two doors meet.
- Pry up the cut piece.
- Locate the sliding bolt and remove it.

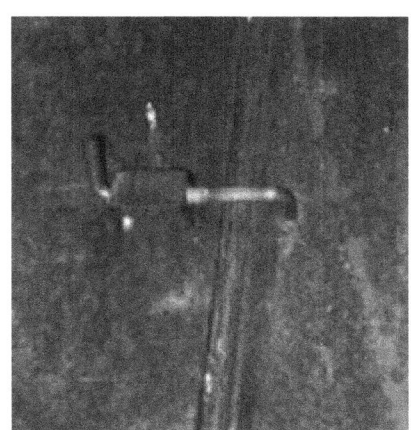

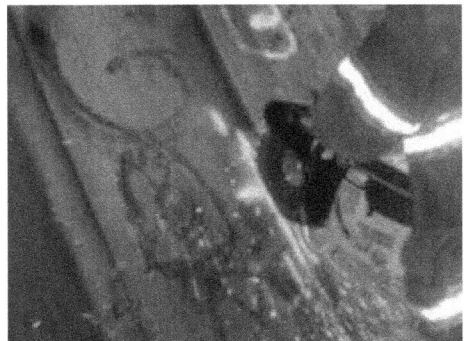
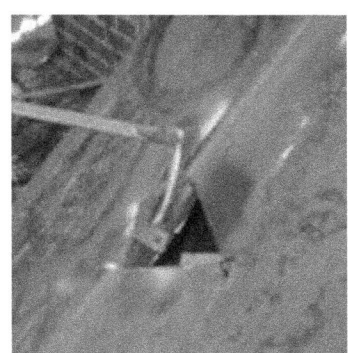

BULKHEAD DOORS

Security of bulkhead doors can vary. Some may be quite formidable and others may be secured by a simple locking device (sliding bolt or hook and eye). The simple device is usually not a problem.

It is important to open this door because victims may attempt to flee the fire by going up to the roof.

BULKHEAD DOORS

Chain and Padlock (Formidable Device)

A hole cut through the door and the bulkhead and secured with a chain. There is nothing to get leverage on, since the chain will slip through the hole.
In this instance, **attack the hinge side first** and pivot the door on the lock side.

An **alternate method** would be to batter the door at the chain attempting to break the door at the opening holding the chain.

Note: If there is a heavy smoke condition, attempt to vent the skylight above the bulkhead first. This will vent the stairs while you are attempting to open the door.

Tight Door in a Metal Frame

Drive the **ADZ** of the Halligan Tool into the side, top or bottom of the door, to get a purchase point. Another method would be to use the **PIKE** end. By "toeing" the Halligan Hook, you can use the hook to drive the Halligan into the seam to get a purchase. Work the Halligan Tool down to the area of the locking device and force the door.

Note: There is usually some clearance at the top and/or bottom of most doors.

BULKHEAD DOORS

Tight Door in a Metal Frame

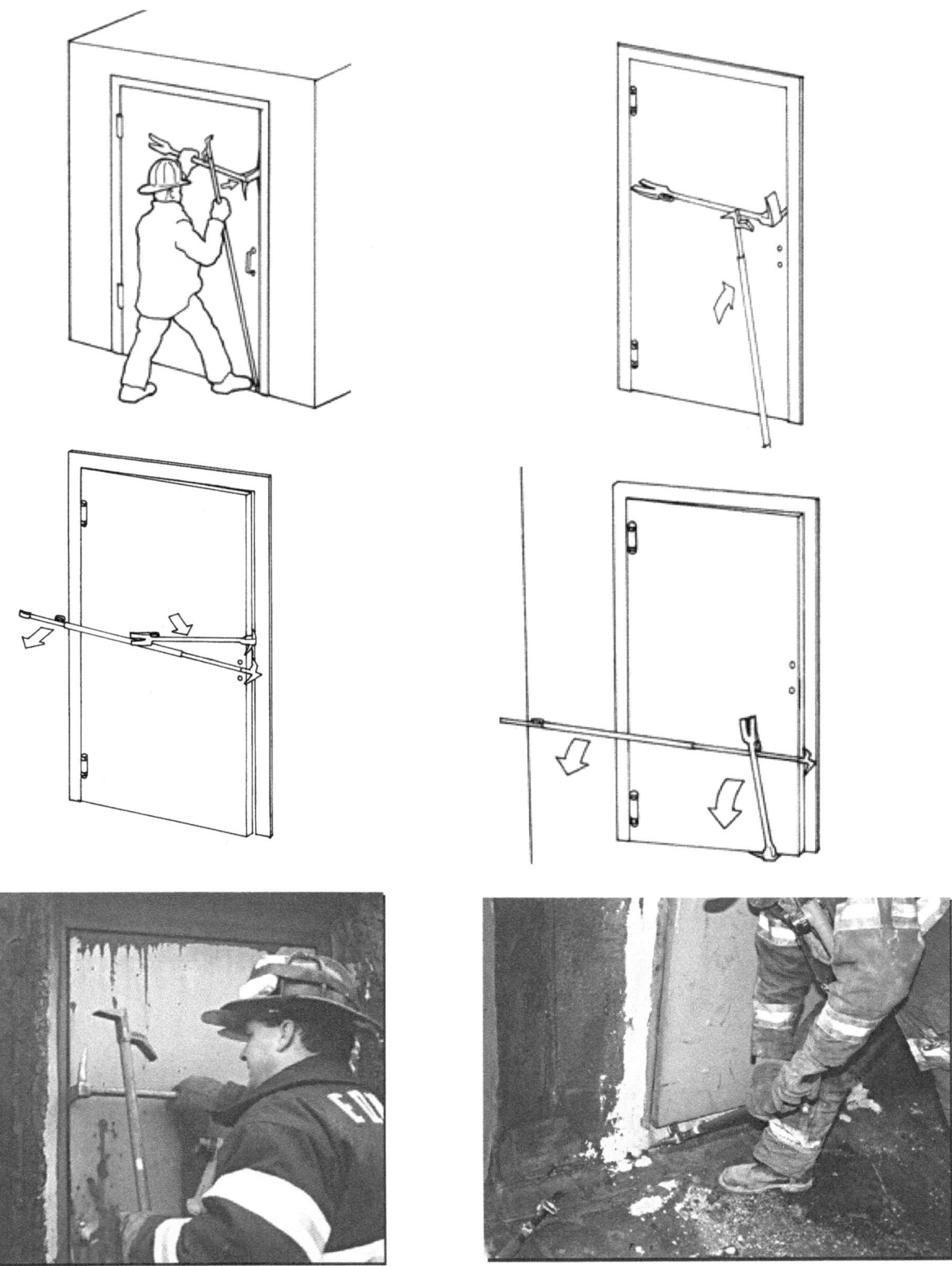

TIPS AND TECHNIQUES

TIPS

Halligan Tool
Marking the Halligan Tool for judging depth of the door when setting the tool.

Mark the adz and the fork with a notch approximately 1¾ inches up from the end to denote the depth of the door. When trying to "lock" the tool in, this will give you assurance that you are in deep enough.

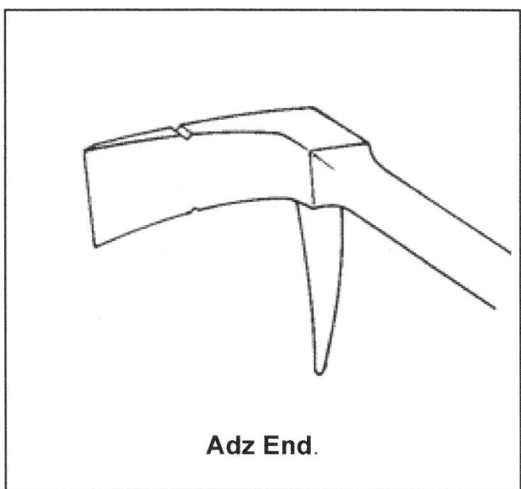

Adz End.

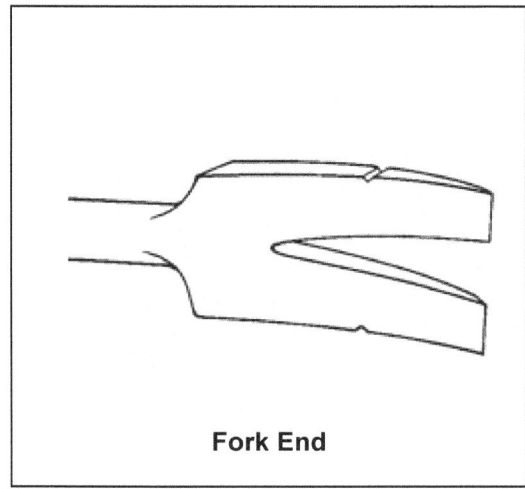

Fork End

Squaring the Shoulder
By squaring the shoulder of the fork end, it allows another striking surface in tight spaces and when the tool is recessed.

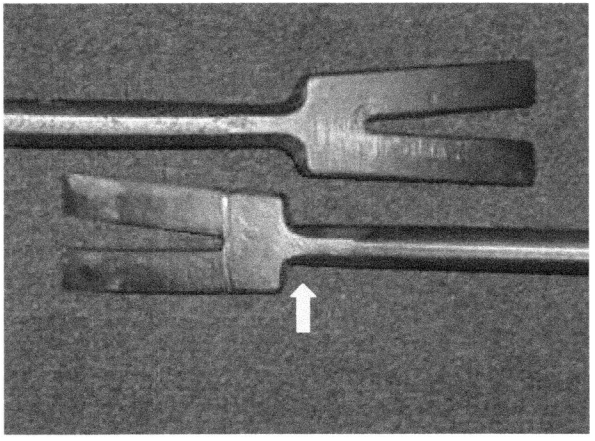

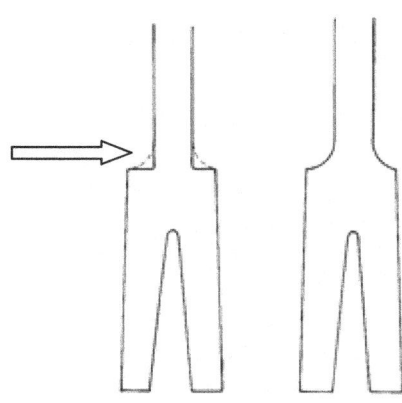

TIPS

Notching the Adz
This may give you the option of pulling a cylinder, getting a lock on, or shearing a bolt or screw. The edge is filed to a clean edge.

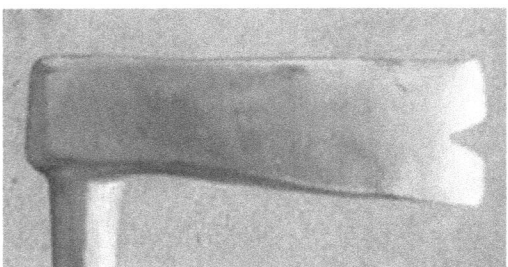 

Notching the Axe Blade
This will give you the ability to "marry" the axe to the Halligan and carry both tools in one hand.

If this is done with a file, there is little chance of damaging the tool.

Usually this is not necessary with a 6 lb. axe, but with an 8 lb. axe the head is too heavy for the fork to slip over.

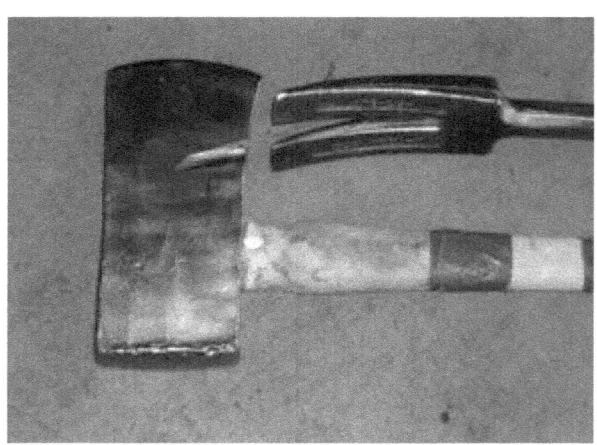

Modifying The Handle of The Halligan Hook

The butt end of the handle is quite useful as a PRYING TOOL or as a SHUT-OFF KEY.

TIPS

Loop In Halligan Hook For A Chain

This simple device allows another tool to be attached for breaking windows beyond the reach of just the hook.

A similar device must also be attached to the other tool, e.g. Halligan Tool.

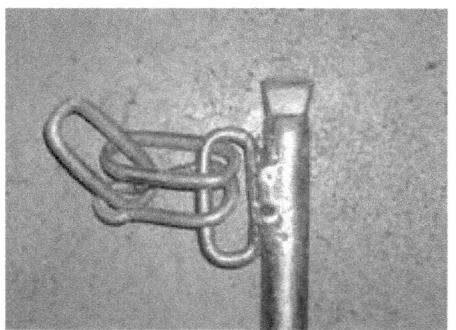
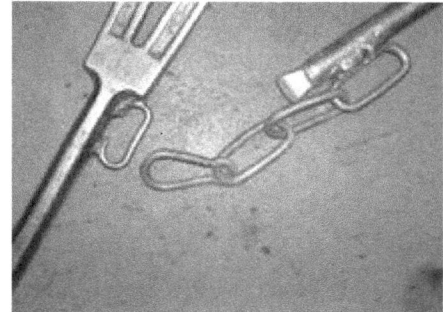

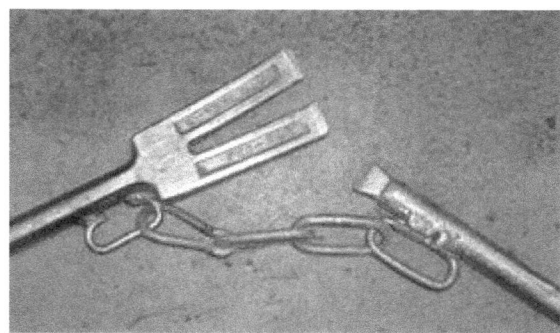

GLOSSARY OF TERMS

"A" Tool	A lock puller (Officer's Tool).
Adz	The axe-like tool with a curved blade at right angle to the handle (shaft).
Arch	The inside curve on the fork end of the Halligan Tool where the two blades of the fork are joined.
Bam-Bam Tool	A dent puller adapted for pulling lock cylinders.
Batter the Door,	Striking the door, doorframe, with an axe, maul or Halligan Tool.
Bevel Side	The curved side of the fork end of the Halligan Tool.
Bolt	A fastening device that is square or round that slides into a notch (keeper).
Chocking the Door	A means of keeping a door open.
Claw Tool	A forcible entry tool.
Cylinder Guard	A metal security plate mounted over the lock cylinder.
Door Flexes	Door bends but does not break or open.
Door Frame	A structural boarder into which a door is hung, also known as a doorbuck or doorjamb.
Door Rail	The outer edge of a door, usually the strongest part of the door.
Door Stop	That portion of the doorframe that prevents the door from winging past the frame.
Doughnut Lock	American 2000 series lock "Hockey Puck."
Duckbill Lock Breaker	A tool designed for forcing padlocks.
Fasco Lock	American 2000 series lock used for securing maintenance rooms in public housing developments.
Fox Lock	A double bar lock.
Gap the Door	The initial opening made in the door and or frame to create a purchase point.
Glass Door	A tempered glass door.
Guarded Keyway	A device over the keyway to prevent the keyway from being removed.
Halligan Tool	Forcible entry tool.
Hockey Puck	American 2000 series lock "Doughnut Lock."

Horseshoe Padlock	Type of heavy-duty lock.
HUD Window/Door	A method of securing an opening with plywood and 2 x 4's horizontal to the opening securing the plywood.
Hydra-Ram	A hydraulic forcible entry tool.
Inward Opening Door	Door swings **AWAY** from you.
Irons	Set of forcible entry tools, usually an axe and Halligan Tool.
Jimmying a Door	Separating (spreading) of the door away from the jamb.
K-Tool	A tool designed for pulling lock cylinders.
Kalameine Door	A door covered with metal.
Kelly Tool	A forcible entry tool.
Key Tool	A set of tools used in conjunction with K-Tool to open locks.
Latch	A fastening device that is angled to slide into a notch (keeper).
Lock In	Getting the Halligan Tool or lock puller tool behind the doorframe.
Lock Puller	A tool designed to pull lock cylinders.
Mortise Lock	A locking device that is designed to fit into the cavity in the edge of the door.
Multi-Lock Door	A door with an integrated lock system which has four pins locking the door into the jamb at four different points.
Officer's Tool	`Lock puller or "A" tool.
Outward Opening Door	Door swings **TOWARD** you.
Pivoting Deadbolt	A fastening device that is square and pivots into a notch (keeper).
Pocket Door	An interior sliding door that slides into a partition or cavity in the wall.
Police Lock	A vertical bar lock.
Purchase Point	The opening made in the door / door frame for forcing the door.
Rabbit Tool	A hydraulic forcible entry tool.
Rail of the Door	The outer edge of a door, usually the strongest part of the door.
Replacement Door	A pre-hung door and jamb installed into an **EXISTING** frame.

Rim Lock	A surface mounted lock.
Set the Tool	Driving the Halligan Tool into the GAP until the arch of the fork is even with the door and or the door stop.
Shoulder	The topside of the fork end at the shaft.
Slipping the Door	Moving the Halligan Tool up and down to free the tool that may be stuck.
Springing the Door	Moving the Halligan Tool side to side (in and out) to free the tool that may be stuck.
Stacked Locks	A series of locks placed close together on a door.
Static Bar	A fastening device which can be mounted across the door.
Stem	On a rim lock, that portion of the lock cylinder that locks or unlocks the mechanism.
Thru-the-Lock	Gaining entry by attacking the locking device and opening the door with little or no damage to the door and or frame.
Sunilla Tool	A lock puller.
Tubular Dead Bolt	A cylinder lock that is a cross between a mortise lock, rim lock and a Key-in-the-Knob lock. May be double-keyed.
Warehouse Window/Door	A method of securing an opening with plywood and 2x4's angled to the floor.
Wrapped Lock	Padlock with steel welded to the lock.